Self-Complementary Antennas

Principle of Self-Complementarity for Constant Impedance

Yasuto Mushiake

Self-Complementary Antennas

Principle of Self-Complementarity for Constant Impedance

With 93 Figures

Springer

Yasuto Mushiake, B.Eng, PhD

Tohoku University and Tohoku Institute of Technology
2–18 Akebonomachi
Aobaku
Sendai 981
JAPAN

ISBN-13: 978-1-4471-1255-6 e-ISBN-13: 978-1-4471-1003-3
DOI: 10.1007/978-1-4471-1003-3
Springer-Verlag Berlin Heidelberg New York

British Library Cataloguing in Publication Data
Mushiake, Yasuto
Self-complementary antennas : principle of self-complementarity for constant impedance
1. Antennas (Electronics) 2. Radio – Antennas 3. Impedance (Electricity)
I. Title
621.3′824

Library of Congress Cataloging-in-Publication Data
Mushiake, Yasuto, 1921-
Self-complementary antennas : principle of self-complementarity for constant impedance/Yasuto Mushiake.
p. cm.
Includes bibliographical references and index.

1. Adaptive antennas. 2. Impedance matching. I. Title.
TK7871.6.M87 1996
621.384′135—dc20 95-25745

Typesetting: Keyword Typesetting Services Ltd., Wallington, Surrey

69/3830-543210 Printed on acid-free paper

Contents

Preface ix

Acknowledgements xiii

1 Introduction 1

1.1 Self-complementary and related broad-band antennas 1

1.2 Background to the emergence of the self-complementary antenna 2

1.3 Brief history of self-complementary antennas. . . . 3

2 Fundamental Theories of Complementary Structures . . 7

2.1 A pair of mutually dual structures 7

2.2 Symmetrical and anti-symmetrical electromagnetic fields 9

2.3 Electromagnetic fields for complementary planar structures. 13

3 Impedance Relationships for Complementary Planar Structures 15

3.1 Input impedances of mutually complementary two-terminal planar structures. 15

3.2 Input impedances of mutually complementary multi-element planar structures 17

3.3 Examples of complementary planar structures . . . 21

4 Origination of Self-Complementary Planar Structures and Discovery of their Constant-Impedance Property. 25

4.1 Origination of self-complementary planar structures. 25

4.2 Constant-impedance property of self-complementary planar structures 26

4.3 Examples of self-complementary planar structures . . 27

5 Multi-Terminal Self-Complementary Planar Structures 31

5.1 Rotationally symmetric multi-terminal self-complementary planar structures 31
5.2 Single-phase excitations for rotationally symmetric multi-terminal self-complementary planar structures . 35
5.3 Axially symmetric multi-terminal self-complementary planar structures 37
5.4 Axially symmetric two-port self-complementary planar structures. 40
5.5 Coupling-less property between loaded unipole-notch type self-complementary planar structures. 42

6 Three-Dimensional Self-Complementary Structures . . 49

6.1 A pair of dual structures consisting of crossed infinite planar sheets of compound perfect conductors 49
6.2 Three-dimensional complementary structures. . . . 50
6.3 Three-dimensional self-complementary structures . . 51
6.4 Examples of three-dimensional self-complementary structures. 54

7 Stacked Self-Complementary Antennas. 57

7.1 Stacking of self-complementary antennas 57
7.2 Co-planar stacked self-complementary antennas. . . 58
7.3 Some variations of co-planar stacked self-complementary structures 60
7.4 Side-by-side stacked self-complementary antennas . . 63
7.5 Compound-stacked self-complementary antennas . . 67

8 General Considerations about Approximations and Modifications of Self-Complementary Antennas . . . 71

8.1 Approximations and modifications for practical purposes 71
8.2 Approximation by truncation 71
8.3 Approximation by replacement with conducting rods 72
8.4 Modification by deformation 75
8.5 Modification by partial excision 75
8.6 An example of transformation from a self-complementary sheet structure to the conducting rod structure 75

9 Developmental Studies of Rotationally Symmetric Self-Complementary Antennas 81

9.1 Alternate-leaves type self-complementary antenna . . 81
9.2 Approximation and modification of alternate-leaves type self-complementary antennas 86
9.3 Modified four-terminal self-complementary antenna on conical surface 88
9.4 Non-constant-impedance property of incorrectly arranged log-periodic structures 93
9.5 Other developmental studies for derivatives of rotationally symmetric self-complementary structures 96

10 Developmental Studies of Axially Symmetric Self-Complementary Antennas 99

10.1 Experimental study of equally spaced unipole-notch array antenna 99
10.2 Equally spaced unipole-notch array antenna on a conical ground plane for the circularly polarized wave 101
10.3 Unipole-notch alternate array antennas stacked on both edges of an angular conducting sheet 103
10.4 Unipole-notch array antennas formed on the substrate of a printed circuit 108

11 Monopole-Slot Type Modified Self-Complementary Antennas 111

11.1 Monopole-slot type antennas derived from the three-dimensional self-complementary antenna . . . 111
11.2 A monopole-slot antenna element as a limiting case 113

12 Conclusion 119

References 121

Subject Index 125

Author Index 127

Preface

Remarkable developments have been made in the utilization of radio waves, and a number of antennas with various characteristics designed for specific systems are being used. Those antennas are essential components as radiators, not simply for the radio-wave utilization systems themselves but also for a variety of other systems related to associated test equipment, radio environmental measuring equipment, and so on. In such circumstances, the significance of antenna technology is becoming increasingly important.

For the antennas used in these systems, an efficient transducing capability of the electric power between the electric circuits and the electromagnetic waves is required for the assigned frequencies. Moreover, such a capability is very often necessary for broad-band frequencies also. Accordingly, broad-band characteristics of input impedances and power gains are desirable for practical antennas that are connected to transmission lines.

On this account, the realization of broad-band frequency characteristics for antennas has been a constant endeavour of the engineers and scientists working in this field, and numerous attempts have been made to this end since research began on the subject. Among the historic events in the development of the broad-band antennas, the discovery of the *Principle of self-complementarity*, which realizes constant-impedance structures, can be regarded as one of the most distinctive events, because this principle provided the origin of frequency-independent antennas, subsequently developed to the so-called log-periodic antennas and log-periodic dipole array. Although the names for these antennas were inadequate, this type of modified self-complementary antenna and its derivatives are still being used extensively in practice as extremely broad-band practical antennas.

The above-mentioned principle, discovered in 1948, created a breakthrough in antenna evolution by providing theoretically constant-impedance structures for antennas, and various extremely broad-band antennas have been developed under its application. Also, viewed from the standpoint of electromagnetic wave theory, the principle created an entirely new concept for structures with conducting planes and excitation sources. Furthermore, extension of the principle to various other cases can be anticipated, even to cases other than those of antennas. However, the technical value of the self-complementary structures and the theoretical significance of this principle have not been properly recognized until quite recently.

For this reason, it has been a long-cherished plan of the present author to publish a compact monograph that describes the comprehensive results of studies based on the Principle of self-complementarity and its application to extremely broad-band antennas. Thus the present text came into existence.

In this book, the fundamental theories related to this principle are first rigorously treated, and then the origination of self-complementary planar structures and the discovery of their constant-impedance property are briefly explained. Next, extensions of the principle to various other cases are discussed theoretically.

In the latter half of the book, the results of developmental studies are described, approximations and deformations from the original structures derived by the theory are discussed, and the application of the theory to extremely broad-band practical antennas is explained. Many experimental data are given as examples of the results obtained from experimental studies.

To aid better understanding for readers, a number of figures and tables have been introduced into the text. This is because structural shapes are especially important for self-complementary structures. Furthermore, practically useful fundamental data for designing extremely broad-band antennas are included in these figures and tables.

In this book, however, in general results of archival importance related to the subject are described by placing particular emphasis on those results obtained by the author himself and the associated group. Related papers and other reports are listed in the References located at the end of the book. The author earnestly hopes that this short book will have significant impact upon the developments of science and technology, not only in the field of antennas and electromagnetic waves but also in other related fields.

The plan to publish this book was encouraged and supported by Professor Chen-To Tai, University of Michigan, and Dr W. Ross Stone, Editor-in-Chief, *IEEE Antennas and Propagation Magazine*. Also, in the course of arranging this manuscript, the advice of Emeritus Professor Saburo Adachi and Professor Kunio Sawaya, Tohoku University, was very helpful, and suggestions from Professor Akira Ishimaru, University of Washington, Emeritus Professor Toshio Sekiguchi, Tokyo Institute of Technology, Dr Kiyoshi Nagai, Toshiba Corporation, Professor Hisamatsu Nakano, Hosei University, and Dr Ken-ichi Kagoshima, Nippon Telegraph and Telephone Corporation, were really valuable.

Publication of this volume has been made possible through the generous co-operation of the Springer-Verlag group, especially helpful being the thoughtful assistance of Mr Nicholas Pinfield, Engineering

Editor of Springer-Verlag London Ltd, and the weighty support of Professor Peter J.B. Clarricoats, University of London.

The author would like to take this opportunity to express his sincere gratitude to all the persons mentioned above for their warm-hearted courtesy.

At Sendai, May 1995 *Yasuto Mushiake*

Acknowledgements

The author wishes to express his sincere gratitude to the late Professor Shintaro Uda for kind advice and pertinent directions over a long period of time extending back to the start of the author's graduate study on Yagi–Uda antenna at Tohoku Imperial University.

He also wishes to thank Professors Saburo Adachi and Tsukasa Yoneyama for their helpful discussions and kind advice.

Furthermore, the author sincerely wishes to thank Professor Takayuki Ishizone and Dr Takeshi Kasahara for their truly helpful co-operation during the intensive developmental studies of these antennas, and for their outstanding contributions, as described in the text.

Grateful acknowlegement is also due to the following investigators for their excellent contributions; Professors Norio Nasu, Kazuhito Matsumura, Naoki Inagaki, Kunio Sawaya, Susumu Horiguchi, Tsuneo Furuya, and other members of this research group who formerly studied at Mushiake Laboratory, Tohoku University, including Dr Masaaki Kudo, Katsuhiro Yamamoto, Shoichi Nishimura, Yukio Yokoyama and Hitoshi Ishikawa.

Also the author wishes to take this opportunity of expressing his sincere and deep thanks to the late Most Reverend Shozen Nakayama, and to the deceased parents of the author, Kin-ichi and Yakuno Mushiake, for their foresighted understanding of his educational needs.

It must also be mentioned here that warmest encouragement and helpful advice was given unstintingly to the author by his wife Michiyo, daughter Kumiko, son Hajime, and son's wife Miki. Sufficient thanks for all this devotion cannot adequately be expressed.

---

1 • INTRODUCTION

1.1 Self-complementary and related broad-band antennas

An antenna with a self-complementary structure has a constant input impedance, independently of the source frequency and shape of the structure. The practical realization of such a structure, and an explanation of the remarkable principle of constant impedance, were accomplished by Yasuto Mushiake [1.1–1.3] in 1948. Various other types of self-complementary structures were then developed successively over a fairly long period of time until quite recently [1.4–1.9].

The principle of self-complementary antennas was also introduced into the investigation of extremely broad-band, or frequency-independent antennas, proposed by V.H.Rumsey [1.10, 1.11], and lead to the development of the log-periodic antenna [1.12] and the log-periodic dipole array [1.13]. These developmental studies in the United States concerned just one basic type of self-complementary structure. Further intensive developmental studies were carried out at Tohoku University and its related institutions in Japan, to investigate other types of self-complementary antenna.

In this connection, it should be noted that the log-periodic structure does not automatically ensure frequency-independent characteristics for antennas, as explained in *The New IEEE Standard Dictionary of Electrical and Electronics Terms*,[1] although self-complementary structures and their modifications do guarantee the constant impedance of an antenna. This means that the broad-band characteristics of the log-periodic antenna have their origin not in the log-periodic shape, but rather in those aspects of the shape that are derived from the self-complementary antenna.[2]

However, since the self-complementary property of an antenna prescribes nothing about the broad-band nature of its radiation properties, information about the radiation characteristics of other existing antennas must be

[1] *Log-periodic antenna* Any one of a class of antennas having a structural geometry such that its impedance and radiation characteristics repeat periodically as the logarithm of frequency.

[2] It was confirmed that an antenna with a log-periodic structure arranged in an anti-complementary manner has a log-periodic input impedance which varies distinctly with respect to the source frequency [1.6]. This fact suggests that the log-periodic dipole array, excited by a feed line without transposition, does not have good broad-band characteristics.

considered if practical broad-band antennas are to be developed. In addition, self-complementary antennas have infinitely extended structures, and reduction of the effects of truncation is very important in practice. Experimental "hard" studies are therefore usually needed for this purpose.

Properly arranged log-periodic structures have various well-known merits as extremely broad-band antennas, but this is not so for such badly arranged structures. Reduction of truncation effects, mentioned above, can be achieved satisfactorily for log-periodic structures, as for other cases of "teeth-type" array structures in general.

In this book, a series of studies on self-complementary antennas will be described in detail, with particular emphasis being placed on the results of original work performed by the present author.

The theory of the self-complementary antenna provides a firm foundation and effective guidance principles for technological investigations of extremely broad-band antennas.

1.2 Background to the emergence of the self-complementary antenna

During the early 1940s, the vital importance of radio equipment was recognized, and investigations in related fields, especially concerning antennas, attracted considerable attention in Japan. Above all, studies on sharp-beam and low-profile antennas, effective in the very-high-frequency region, were considered to be especially important and to merit urgent investigation.

In such circumstances, the present author started his graduate study at Tohoku Imperial University, in 1944, under the direction of Professor Shintaro Uda, who was the successor to Professor Hidetsugu Yagi at the same university.

Initially, this research was concerned with the theoretical method of designing Yagi–Uda antenna [1.14]. In that connection, the determination of current distributions along antenna conductors is very important. Hallén's theory [1.15] was therefore introduced to resolve this problem, based on the boundary conditions on the surfaces of the conductors.

However, the slit[3] antenna had already been proposed in Japan as an ideal low-profile antenna on a conducting surface. The author believed that introducing such a slit antenna into the sharp-beam antenna system might be advantageous. However, at that time, the method used to analyse slit antennas was based on the assumed magnetic current distribution over the slit surface. By comparing this situation with that of wire antennas, the author became convinced that a new theory had to be originated for slot antennas, which did not involve any such assumptions.

[3]The slot antenna was called the slit antenna in Japan at that time.

In the meantime, the author had been deeply impressed by *Electromagnetic Theory* by J.A. Stratton [1.16],[4] and he studied this book very carefully during his undergraduate and graduate studies. In addition, some basic studies were done in Japan during the mid-1940s on the application of Babinet's principle to electromagnetic fields.

Under such circumstances, the present author regarded slit antennas as a secondary subject of his investigations, his principal studies being the Yagi–Uda antenna. Unexpectedly, however, study of the secondary subject lead to the emergence of self-complementary antennas.

1.3 Brief history of self-complementary antennas

As a result of the theoretical and fundamental investigations of slot antennas performed by the author, he succeeded in obtaining a general and innovative mathematical relationship between two input impedances for mutually complementary planar structures. This relationship was derived via an exact and general theory, and holds for an arbitrarily shaped planar structure. However, the form of the new relationship was identical with that of the previous one, which had been obtained by several other investigators, for a narrow slot antenna and a complementary wire antenna, after making various assumptions. For this reason, the author's work was disregarded by his colleagues and his superiors, and he was discouraged from making any report on the result.

However, on reconsidering the matter, he realized that, to be taken seriously, he must describe convincing examples of applications that clearly highlighted the distinctive features of the innovative relationship. After intensive studies to this end, an entirely new type of structure was originated by the author, in 1948. This was the self-complementary planar structure having a constant input impedance. Such was the discovery of the *Principle of self-complementarity*.

The first report on this work was presented at a local convention in 1948 [1.1], and the second contributed to the *Journal of the Institute of Electronic Engineers of Japan* (in Japanese), an abstract becoming available in 1949 [1.2]. However, the scheduled six-page text ultimately remained unpublished, because of a paper shortage in Japan just after World War II. The third report [1.3] was hence published as a university report (in English) in 1949. The types of self-complementary structure illustrated in these reports were of two kinds: (1) the balanced type two-terminal structure; and (2) the unbalanced type two-terminal structure.

Around that time, preparations were being made for the inauguration of television broadcasting in Japan. Under such circumstances, the establishment of a reliable design method for Yagi–Uda antenna became urgent, and the

[4]Jurius A. Stratton's letter of 24 November 1947 to Yasuto Mushiake, in response to a letter of 11 October 1947, pointed out certain misprints in the first edition of his book *Electromagnetic Theory*.

author had to concentrate his energy to that end [1.17, 1.18]. Consequently, his studies on self-complementary structures had to be temporarily suspended. However, the importance of such structures was firmly embedded in his mind, and he was always on the look-out for opportunities to resume research into the use of self-complementary structures as antennas.

Meanwhile, the paper published in English in 1949 attracted the attention of V.H. Rumsey in the United States, and the principle of self-complementary structure was introduced into the developmental studies of frequency-independent antennas proposed by him in 1957 [1.10]. Furthermore, related studies were actively performed in the United States.

On learning about these developments, the author realized that a good opportunity might have come to resume his own studies on self-complementary antennas. However, investigations into linear antennas were still being actively continued at Tohoku University [1.19, 1.20], and he therefore resumed that study as a secondary objective, although it progressed very slowly. The results made available in 1959 were an extension of the principle to the case of rotationally symmetric multi-terminal self-complementary antennas [1.4].

In contrast, developments on frequency-independent antennas in the United States were outstanding [1.11]. The developmental studies on log-periodic antennas by R.H. Duhamel and D.H. Isbell [1.12], which lead to the log-periodic dipole arrays by D.H. Isbell [1.13], and the detailed discussions on multi-terminal self-complementary antennas by G.A. Deschamps [1.21], should particularly be emphasized as examples of the excellent achievements in this field.

In Tohoku University, however, studies in this field were continued by the author, although very slowly because it was still only a secondary subject. This research aimed at extending the principle to constant-impedance antennas in general, and the three-dimensional self-complementary antenna was originated in 1963 [1.5].

Around the mid-1970s, T. Ishizone, N. Inagaki, N. Nasu and T. Kasahara joined in developmental studies on self-complementary antennas, which were performed not just in Tohoku University but extended to related institutions, all being under the direction of the present author. The studies performed by this group were supported by financial assistance from the Hoso Bunka Foundation, Japan (1976–77), and Grant-in-Aid for Scientific Research, Ministry of Education, Japan (1980–83). Thus, intensive studies were continued until the mid-1980s. During that period, the strong and earnest co-operation of K. Sawaya and other members of the Mushiake Laboratory, Tohoku University, was also made available.

As a result of the theoretical and developmental studies by this group, several new types of self-complementary antenna were originated, and the effects of approximation and modification were clarified [1.8, 1.22]. These results provided various important information and data essential for the practical development of extremely broad-band antennas. Various types of self-complementary antenna, originated by the author himself, are listed in Table 1.1.

Some supplementary studies were continued until quite recently, and these results are also included in this book.

Table 1.1 Various types of self-complementary antennas (SCA) originated by Yasuto Mushiake

No.	Year	Type of SCA structure*	Input impedance (Ω)
1	1948	Rotationally symmetric 2-terminal planar SCA (balanced type)	60π
2	1948	Axially symmetric 2-terminal planar SCA (unbalanced type)	60π
3	1957	Rotationally symmetric 4-terminal planar SCA (turnstile)	$30\sqrt{2}\pi$ each
4	1963	Three-dimensional multi(N)-planar 2-terminal SCA	$60\pi/N$
5	1982	Co-planar stacked SCA	60π each
6	1982	Side-by-side stacked SCA	60π each
7	1985	Compound-stacked SCA**	60π each

*Restricted to the types of SCA originated personally by Mushiake.

**Japanese Patent No. 1594525, Y. Mushiake, 27 December 1990.

2 • FUNDAMENTAL THEORIES OF COMPLEMENTARY STRUCTURES

2.1 A pair of mutually dual structures

For the purpose of developing the theory of self-complementary antennas, some preliminary considerations will be made in this chapter, and electromagnetic fields for a pair of mutually dual structures will be discussed in this section.

Initially, a structure with the general configuration shown in Fig. 2.1(a) will be considered, where an arbitrarily shaped perfect electric conductor (PEC) and a perfect magnetic conductor (PMC) are used, and arbitrarily distributed electric and magnetic source currents are assumed [2.1]. Next, the structure shown in Fig. 2.1(b) will be considered, where the geometrical structure is identical to that of Fig. 2.1(a), except that the electric and the magnetic properties are mutually interchanged between them.

Then, the electromagnetic fields, $\boldsymbol{E}_1$, $\boldsymbol{H}_1$ for Fig. 2.1(a) and $\boldsymbol{E}_2$, $\boldsymbol{H}_2$ for Fig. 2.1(b), and the electric and magnetic source currents, $\boldsymbol{J}_0$ and $\boldsymbol{J}_{0m}$, satisfy the Maxwell's equations as follows:

$$\left.\begin{aligned} \nabla \times \boldsymbol{E}_1 + \mathrm{j}\omega\mu\boldsymbol{H}_1 &= -\boldsymbol{J}_{0m} = -\boldsymbol{M} \\ \nabla \times \boldsymbol{H}_1 - (\mathrm{j}\omega\varepsilon + \sigma)\boldsymbol{E}_1 &= \boldsymbol{J}_0 = \boldsymbol{N} \end{aligned}\right\} \tag{2.1}$$

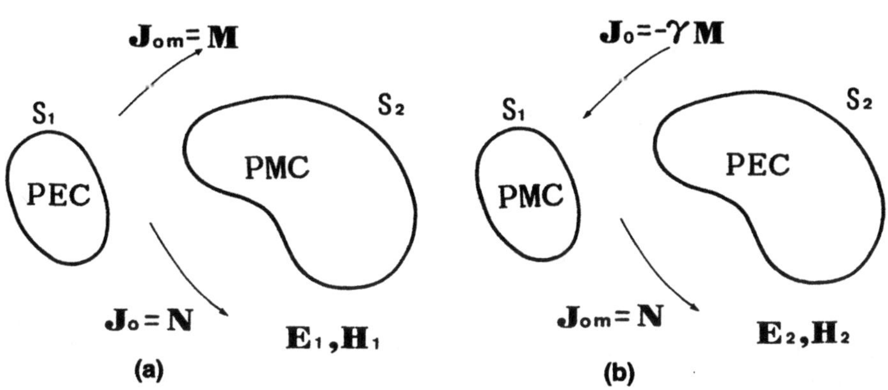

Fig. 2.1 A pair of dual structures.

$$\left.\begin{aligned}\nabla\times\boldsymbol{E}_2+\mathrm{j}\omega\mu\boldsymbol{H}_2&=-\boldsymbol{J}_{0m}=-\boldsymbol{N}\\ \nabla\times\boldsymbol{H}_2-(\mathrm{j}\omega\varepsilon+\sigma)\boldsymbol{E}_2&=\boldsymbol{J}_0=-\gamma\boldsymbol{M}\end{aligned}\right\}\quad(2.2)$$

Here, the MKS rational unit system is adopted, and the sinusoidal time dependence with the factor of exp(jωt) is assumed. The constants ε, μ and σ are permittivity, permeability and conductivity of the medium, respectively, and

$$\gamma=(\mathrm{j}\omega\varepsilon+\sigma)/\mathrm{j}\omega\mu \quad (2.3)$$

Also, the distributions of the electric and magnetic source currents, $\boldsymbol{J}_0$ and $\boldsymbol{J}_{0m}$, are assumed to be $\boldsymbol{N}$ and $\boldsymbol{M}$ in Fig. 2.1(a), and $-\gamma\boldsymbol{M}$ and $\boldsymbol{N}$ in Fig. 2.1(b), respectively.

At the same time, the boundary conditions for the PEC and the PMC must be satisfied on the surfaces S_1 and S_2 in Fig. 2.1(a), and S_2 and S_1 in Fig. 2.1(b), respectively, and we have

$$\left.\begin{aligned}\boldsymbol{E}_1\times\boldsymbol{n}=\boldsymbol{H}_1\cdot\boldsymbol{n}=0&\quad\text{on } S_1\\ \boldsymbol{H}_1\times\boldsymbol{n}=\boldsymbol{E}_1\cdot\boldsymbol{n}=0&\quad\text{on } S_2\end{aligned}\right\}\quad(2.4)$$

$$\left.\begin{aligned}\boldsymbol{E}_2\times\boldsymbol{n}=\boldsymbol{H}_2\cdot\boldsymbol{n}=0&\quad\text{on } S_2\\ \boldsymbol{H}_2\times\boldsymbol{n}=\boldsymbol{E}_2\cdot\boldsymbol{n}=0&\quad\text{on } S_1\end{aligned}\right\}\quad(2.5)$$

where $\boldsymbol{n}$ denotes the unit normal on the surfaces of the perfect conductors.

Now, let the vector symbols $\boldsymbol{E}_2$, $\boldsymbol{H}_2$ in (2.2) and (2.5) be replaced by $\boldsymbol{H}_1$, $\boldsymbol{E}_1$ using the relationships:

$$\boldsymbol{E}_2=-\boldsymbol{H}_1 \quad\text{and}\quad \boldsymbol{H}_2=\gamma\boldsymbol{E}_1 \quad (2.6)$$

Then we have [2.2]

$$\left.\begin{aligned}-\nabla\times\boldsymbol{H}_1+\mathrm{j}\omega\mu\gamma\boldsymbol{E}_1&=-\boldsymbol{N}\\ \gamma\nabla\times\boldsymbol{E}_1+(\mathrm{j}\omega\varepsilon+\sigma)\boldsymbol{H}_1&=-\gamma\boldsymbol{M}\end{aligned}\right\}\quad(2.7)$$

$$\left.\begin{aligned}-\boldsymbol{H}_1\times\boldsymbol{n}=\boldsymbol{E}_1\cdot\boldsymbol{n}=0&\quad\text{on } S_2\\ \boldsymbol{E}_1\times\boldsymbol{n}=-\boldsymbol{H}_1\cdot\boldsymbol{n}=0&\quad\text{on } S_1\end{aligned}\right\}\quad(2.8)$$

By comparing (2.7) with (2.1), and (2.8) with (2.4), it is found that they are respectively identical, except that the order of the description is interchanged between them. This means that the field intensities, $\boldsymbol{E}_1$, $\boldsymbol{H}_1$ and $\boldsymbol{E}_2$, $\boldsymbol{H}_2$, in the relationships are expressed by (2.6) everywhere in space. Consequently, if a solution to either one of Fig. 2.1(a) or 2.1(b) can be obtained, then the solution to the other is immediately given by relationship (2.6). Furthermore, if the magnitudes of the electric and magnetic source currents in Fig. 2.1(b) are varied proportionally by a factor K, then the field intensities

$\boldsymbol{E}_2$, $\boldsymbol{H}_2$ will also be changed proportionally by the same factor K everywhere in space, and we have

$$\boldsymbol{E}_2 = -K\boldsymbol{H}_1 \quad \text{and} \quad \boldsymbol{H}_2 = K\gamma\boldsymbol{E}_1 \tag{2.9}$$

instead of (2.6), where K is an arbitrary factor which could be negative.

Hereafter in this book, a pair of structures such as that shown in Figs 2.1(a) and (b) will be referred to as "mutually dual structures", and relationships (2.6) and (2.9) will be called "duality relations for the electromagnetic fields".

2.2 Symmetrical and anti-symmetrical electromagnetic fields

In this section, the properties of the electromagnetic fields induced by the electric or magnetic source currents that are symmetrically or anti-symmetrically distributed with respect to the plane $x = 0$ of the rectangular coordinates, x, y, z, will be investigated.

Initially, a pair of symmetrical electric source currents, $\boldsymbol{J}_{01}(x', y', z')$ and $\boldsymbol{J}_{02}(-x', y', z')$, as shown in Fig. 2.2, will be considered.

It is easily found from the symmetry that their components are given by the relationships

$$\left.\begin{aligned} \boldsymbol{J}_{01x}(x', y', z') &= -\boldsymbol{J}_{02x}(-x', y', z') \\ \boldsymbol{J}_{01y}(x', y', z') &= \boldsymbol{J}_{02y}(-x', y', z') \\ \boldsymbol{J}_{01z}(x', y', z') &= \boldsymbol{J}_{02z}(-x', y', z') \end{aligned}\right\} \tag{2.10}$$

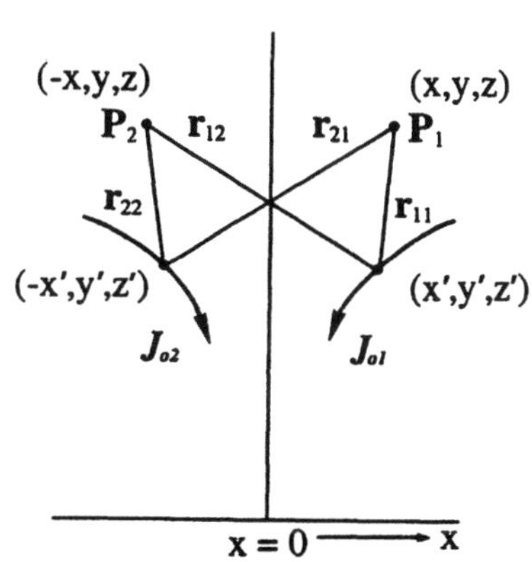

Fig. 2.2 Symmetrical electric source currents.

Hereafter, three-dimensional functions, $f(x,y,z)$ and $f(-x,y,z)$, will be denoted by $f(x)$ and $f(-x)$ for simplicity. Consequently, the Hertzean vector, $\mathbf{\Pi}(x)$, can be expressed as:

$$\mathbf{\Pi}(x) = \mathbf{\Pi}_1(x) + \mathbf{\Pi}_2(x) \tag{2.11}$$

where

$$\left.\begin{aligned} \mathbf{\Pi}_1(x) &= \int \boldsymbol{J}_{01}(x')\phi_{1j}\mathrm{d}v \\ \mathbf{\Pi}_2(x) &= \int \boldsymbol{J}_{02}(x')\phi_{2j}\mathrm{d}v \end{aligned}\right\} \tag{2.12}$$

$$\phi_{ij} = \frac{\exp(-\mathrm{j}kr_{ij})}{\mathrm{j}4\pi\omega\varepsilon r_{ij}} \tag{2.13}$$

$$(k^2 = \omega^2\mu\varepsilon - j\omega\mu\sigma)$$

and r_{ij} is the distance between the two points $\boldsymbol{J}_{0i}(x')$ and $P_j(x)$, which is given as in Fig. 2.2 by

$$r_{ij}^2 = (x'-x)^2 + (y'-y)^2 + (z'-z)^2 \qquad (i,j = 1,2) \tag{2.14}$$

By introducing relationships (2.10) into (2.12), $\Pi_{1x}(P_2)$ and $\Pi_{2x}(P_2)$ can be transformed as follows:

$$\begin{aligned} \Pi_{1x}(-x) &= \int J_{01x}(x')\phi_{12}\mathrm{d}v \\ &= -\int J_{02x}(-x')\phi_{21}\mathrm{d}v \\ &= -\Pi_{2x}(x) \end{aligned}$$

$$\begin{aligned} \Pi_{2x}(-x) &= \int J_{02x}(-x')\phi_{22}\mathrm{d}v \\ &= -\int J_{01x}(x')\phi_{11}\mathrm{d}v \\ &= -\Pi_{1x}(x) \end{aligned}$$

Hence we have from (2.11)

$$\Pi_x(P_2) = -\Pi_x(P_1) \tag{2.15}$$

Likewise, similar transformations can be made as follows:

$$\begin{aligned} \Pi_{1y}(-x) &= \int J_{01y}(-x')\phi_{12}\mathrm{d}v \\ &= \int J_{02y}(x')\phi_{21}\mathrm{d}v \\ &= \Pi_{2y}(x) \end{aligned}$$

$$\begin{aligned}\Pi_{1z}(-x) &= \int J_{01z}(-x')\phi_{12}\mathrm{d}v \\ &= \int J_{02z}(x')\phi_{21}\mathrm{d}v \\ &= \Pi_{2z}(x)\end{aligned}$$

Hence we have from (2.11)

$$\left.\begin{aligned}\Pi_y(P_2) &= \Pi_y(P_1) \\ \Pi_z(P_2) &= \Pi_z(P_1)\end{aligned}\right\} \tag{2.16}$$

Now, the electromagnetic fields can be calculated from $\boldsymbol{\Pi}$ as

$$\left.\begin{aligned}\boldsymbol{E} &= \nabla\nabla\cdot\boldsymbol{\Pi} + k^2\boldsymbol{\Pi} \\ \boldsymbol{H} &= \mathrm{j}\omega\varepsilon\nabla\times\boldsymbol{\Pi}\end{aligned}\right\} \tag{2.17}$$

By introducing relationships (2.15) and (2.16) into (2.17) and performing some mathematical manipulations, we obtain:

$$\left.\begin{aligned}E_x(-x) &= -E_x(x) \qquad & H_x(-x) &= H_x(x) \\ E_y(-x) &= E_y(x) \qquad & H_y(-x) &= -H_y(x) \\ E_z(-x) &= E_z(x) \qquad & H_z(-x) &= -H_z(x)\end{aligned}\right\} \tag{2.18}$$

These relationships show that the electromagnetic fields induced by a pair of symmetrical electric source currents are electrically symmetrical with respect to the plane $x = 0$. However, it is found that the same fields are magnetically anti-symmetrical with respect to the same plane. Furthermore, by interchanging the positive and negative signs for three components of $\boldsymbol{J}_{02}(-x)$ in relationship (2.10), it is easily found that the electromagnetic fields induced by a pair of anti-symmetrical electric source currents are electrically anti-symmetrical but magnetically symmetrical with respect to the plane $x = 0$.

Similarly, it can easily be shown that the fields induced by the anti-symmetrical magnetic source currents are magnetically anti-symmetrical but electrically symmetrical. Accordingly, compound sources of symmetrical electric currents and anti-symmetrical magnetic currents also induce electrically symmetrical fields.

Next, let us consider a case where a pair of symmetrical structures, PEC-1 and PEC-2, are placed in the electromagnetic fields excited by a pair of symmetrical electric source currents, as shown in Fig. 2.3.

The primary electromagnetic fields in the present case are symmetrical, as explained above, and hence the induced electric currents over the surfaces of the symmetrical PEC structures must be symmetrical with respect to the reference plane. Consequently, it is concluded that the total electromagnetic

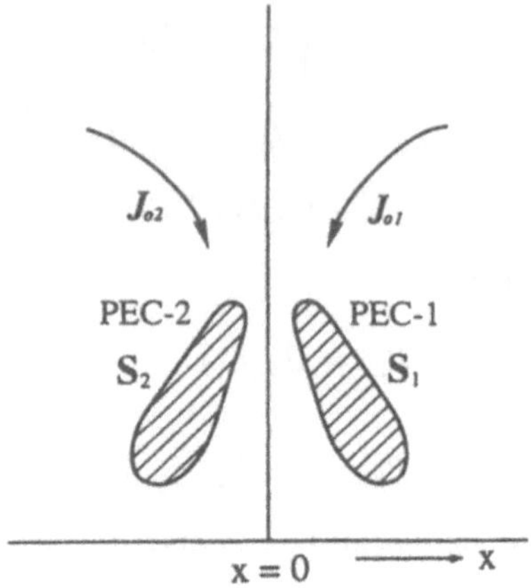

Fig. 2.3 Symmetrical electric source currents with symmetrical perfect electric conductors.

fields, including the secondary fields scattered from the PEC structures, are also symmetrical. In this case, the structures of the PEC could be a single body with symmetrical shape or any portion of a planar PEC sheet inserted on the reference plane. Furthermore, such a symmetrical property is retained even when the symmetrical electric source currents in Fig. 2.3 are replaced by anti-symmetrical magnetic source currents or a combination of symmetrical electric source currents and anti-symmetrical magnetic source currents.

Next, the properties of the symmetrical and the anti-symmetrical electromagnetic fields on the plane of reference will be examined. According to relationships (2.18), the E_x-component of the electrically symmetrical fields is anti-symmetrical with respect to the reference plane $x = 0$ and it is also continuous, and hence we have

$$E_x(-0) = -E_x(+0) = 0 \tag{2.19}$$

or

$$\boldsymbol{E} \cdot \boldsymbol{n} = 0 \quad \text{on } x = 0 \tag{2.20}$$

where $\boldsymbol{n}$ is the unit normal on the plane $x = 0$. At the same time, from (2.18) we find

$$\left. \begin{aligned} H_y(-0) = -H_y(+0) = 0 \\ H_z(-0) = -H_z(+0) = 0 \end{aligned} \right\} \tag{2.21}$$

or

$$\boldsymbol{H} \times \boldsymbol{n} = 0 \quad \text{on } x = 0 \tag{2.22}$$

In addition, magnetically anti-symmetrical fields can be expressed also by (2.18), hence from (2.22) and (2.20) we have

$$H \times n = E \cdot n = 0 \quad \text{on } x = 0 \tag{2.23}$$

In conclusion, the electromagnetic fields induced by the symmetrically distributed electric source currents or anti-symmetrical magnetic source currents in the presence of symmetrical PEC structures have the properties given by (2.23), except within the conductors and on their surfaces.

2.3 Electromagnetic fields for complementary planar structures

If a planar sheet structure with infinite extent is divided into two arbitrary portions, S_1 and S_2, then these two planar structures S_1 and S_2 are complementary to each other. Either one may be of finite extent, or both may be of semi-infinite extent.

In this section, the electromagnetic fields for mutually complementary PEC planar sheet structures excited with symmetrical electric source currents or anti-symmetrical magnetic source currents will be investigated.

To this end, a pair of mutually dual structures formed of PEC and PMC planar sheets, as shown in Figs 2.4(a) and (b) will initially be considered, where S_1 and S_2 are arbitrarily shaped, mutually complementary structures formed of different kinds of perfect conductor. However, their electric and magnetic properties are interchanged between (a) and (b), as indicated in the figure. Besides, a pair of symmetrical electric source currents, N_+, N_-, are assumed in Fig. 2.4(a), while anti-symmetrical magnetic source currents, N_+, $-N_-$, are assumed in Fig. 2.4(b).

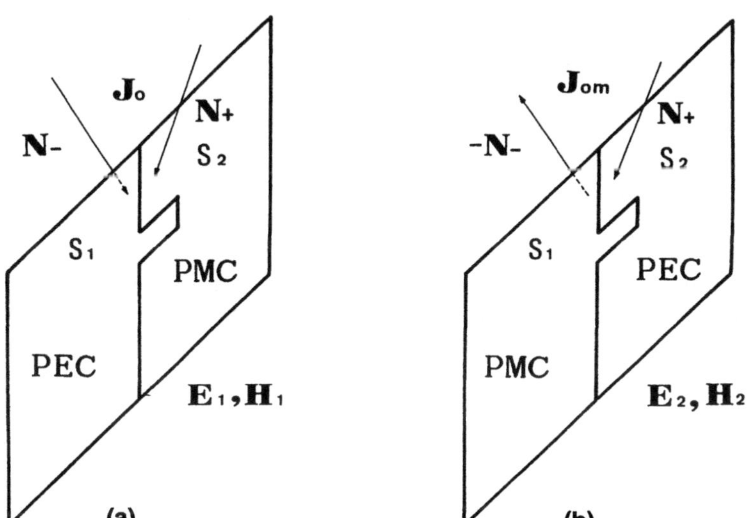

Fig. 2.4 A pair of infinite planar sheets of compound perfect conductors with source currents.

In these two cases, the planes of the perfect conductors, $S_1 + S_2$, extend to infinity, and the whole space in (a) and that in (b) are respectively divided into two independent half-spaces. Therefore, the two half-spaces on the right-hand sides of structures (a) and (b) are mutually dual, and the duality relationships (2.9) hold for the electromagnetic fields $\boldsymbol{E}_1$, $\boldsymbol{H}_1$ and $\boldsymbol{E}_2$, $\boldsymbol{H}_2$ in the two half-spaces. Also with the same argument, similar relationships (2.9) hold for the two half-spaces on the left-hand sides of structures (a) and (b). However, it should be noted here that the directions of the source currents on the left-hand sides are in the opposite direction, while those on the right-hand sides are in the same direction. Therefore, the value of the factor K is -1 for the half-spaces on the left-hand sides of (a) and (b), but $+1$ for the right-hand sides of (a) and (b). Consequently, the duality relationships for the electromagnetic fields (2.9) reduce to

$$\boldsymbol{E}_2 = \mp \boldsymbol{H}_1 \quad \text{and} \quad \boldsymbol{H}_2 = \pm \gamma \boldsymbol{E}_1 \tag{2.24}$$

where the upper and the lower signs apply to the half-spaces on the right-hand side and the left-hand side of the conducting planes, $S_1 + S_2$, respectively.

In addition, a pair of symmetrical electric source currents induce electrically symmetrical electromagnetic fields. However, in the presence of the arbitrarily shaped PEC planar sheet, S_1, relationships (2.20) and (2.22) do not hold on S_1, owing to the effects of the discontinuity of the fields. Hence, we have

$$\boldsymbol{E}_1 \cdot \boldsymbol{n} = \boldsymbol{H}_1 \times \boldsymbol{n} = 0 \quad \text{on } S_2 \tag{2.25}$$

These relationships show that the boundary conditions for the PMC are automatically satisfied on S_2 in Fig. 2.4(a). Also, the anti-symmetrical magnetic source currents induce magnetically anti-symmetrical fields, and in the presence of the arbitrarily shaped PEC planar sheet, S_2, we find

$$\boldsymbol{H}_2 \times \boldsymbol{n} = \boldsymbol{E}_2 \cdot \boldsymbol{n} = 0 \quad \text{on } S_1 \tag{2.26}$$

These relationships also show that the boundary conditions for the PMC in Fig. 2.4(b) are automatically satisfied on S_1.

The properties shown above mean that the PMC sheets on S_2 in Fig. 2.4(a) and on S_1 in Fig. 2.4(b) can be removed, without causing any variation in the original electromagnetic fields. Therefore, we can conclude that relationship (2.24) is also true for a pair of mutually complementary PEC structures, derived from Figs. 2.4(a) and (b), by eliminating the fictitious PMC sheets. Note that, although in the present theory PMC sheets are assumed initially, such a technique is quite helpful for developing the theory to more complicated cases. This is because the introduction of the PMC sheets, as shown in Fig. 2.4, divides the whole space into two perfectly shielded independent half-spaces.

3 • IMPEDANCE RELATIONSHIPS FOR COMPLEMENTARY PLANAR STRUCTURES

3.1 Input impedances of mutually complementary two-terminal planar structures

For the purpose of investigating the relationship between the input impedances for mutually complementary planar structures, an arbitrarily shaped plate antenna and its complementary hole antenna, as shown in Figs 3.1(a) and (b) are considered.

The electric source current in Fig. 3.1(a) can be regarded as a pair of symmetrical electric source currents whose total magnitude is equal to that of the input current. In contrast, the magnetic source current in Fig. 3.1(b) circulates around the bridging conductor, and it is regarded as a pair of anti-symmetrical

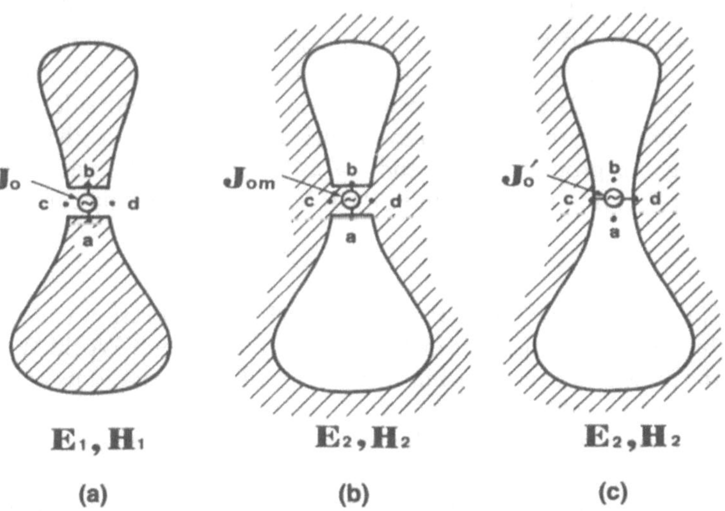

Fig. 3.1 Excitation of mutually complementary planar structures.

magnetic source currents. In addition, the hole antenna in Fig. 3.1(b) is equivalent to that in Fig. 3.1(c) which is excited by an electric source current.

Now, apply the result obtained in Chapter 2, section 2.3 to the electromagnetic fields $\boldsymbol{E}_1$, $\boldsymbol{H}_1$ for the structure in Fig. 3.1(a) and to the electromagnetic fields, $\boldsymbol{E}_2$, $\boldsymbol{H}_2$ for the structure in Fig. 3.1(b) or (c). Then, from the duality relationships (2.24) we have

$$\boldsymbol{E}_2 = \mp\boldsymbol{H}_1 \quad \text{and} \quad \boldsymbol{H}_2 = \pm\gamma\boldsymbol{E}_1 \tag{3.1}$$

under the condition that the magnitudes of the electric and magnetic source currents in Figs 3.1(a) and (b) have the same relationships as those in Figs 2.4(a) and (b). However, the input impedance of the antenna does not depend on the magnitude of the source current, but rather on the ratio of the electric and magnetic field intensities. For this reason, analysis can be continued based on relationship (3.1).

Let the input impedances of the plate antenna and the hole antenna in Figs 3.1(a) and (c) be Z_1 and Z_2, respectively. They can thus be expressed by the ratios of the input voltages and the input currents at the respective feeding terminals as follows:

$$\left.\begin{aligned} Z_1 &= \int_b^a \boldsymbol{E}_1 \cdot \mathrm{d}\boldsymbol{l} \bigg/ 2\int_c^d \boldsymbol{H}_1 \cdot \mathrm{d}\boldsymbol{l} \\ Z_2 &= \int_d^c \boldsymbol{E}_2 \cdot \mathrm{d}\boldsymbol{l} \bigg/ 2\int_b^a \boldsymbol{H}_2 \cdot \mathrm{d}\boldsymbol{l} \end{aligned}\right\} \tag{3.2}$$

where $\mathrm{d}\boldsymbol{l}$ denotes the line element, and the integrations are performed in the vicinity of the input terminals, which are assumed to be sufficiently small compared with the wavelength. The line integrals for the magnetic fields should ideally be performed around the complete paths of the fed currents, but they are expressed in (3.2) as twice the integrations on the front side of the conducting sheets, since they are anti-symmetrical with respect to these sheets.

By taking the product of Z_1 and Z_2 as given by (3.2), and introducing relationship (3.1) into the integrands, the following relationship can be derived:

$$Z_1 Z_2 = 1/4\gamma = (Z_0/2)^2 \tag{3.3}$$

where Z_0 is the intrinsic impedance of the medium. Expression (3.3) had been derived by several investigators in Japan prior to the end of World War II [3.1, 3.2], as the impedance relationship between a slot antenna and its complementary wire antenna, although the results remained unpublished until after the war [see, for example, 3.3, 3.4]. However, it should be stressed here that relationship (3.3) in the author's theory is derived without any restriction being placed on the shape of the antennas, such as whether it is a slot or a wire. Accordingly, this is an innovative and generalized relationship for a pair of arbitrarily shaped, mutually complementary two-terminal planar structures [1.1–1.3].

3.2 Input impedances of mutually complementary multi-element planar structures

As the simplest case of a multi-element planar structure, arbitrarily shaped two-element plate antennas and their complementary hole antennas as shown in Fig. 3.2 will be initially discussed.

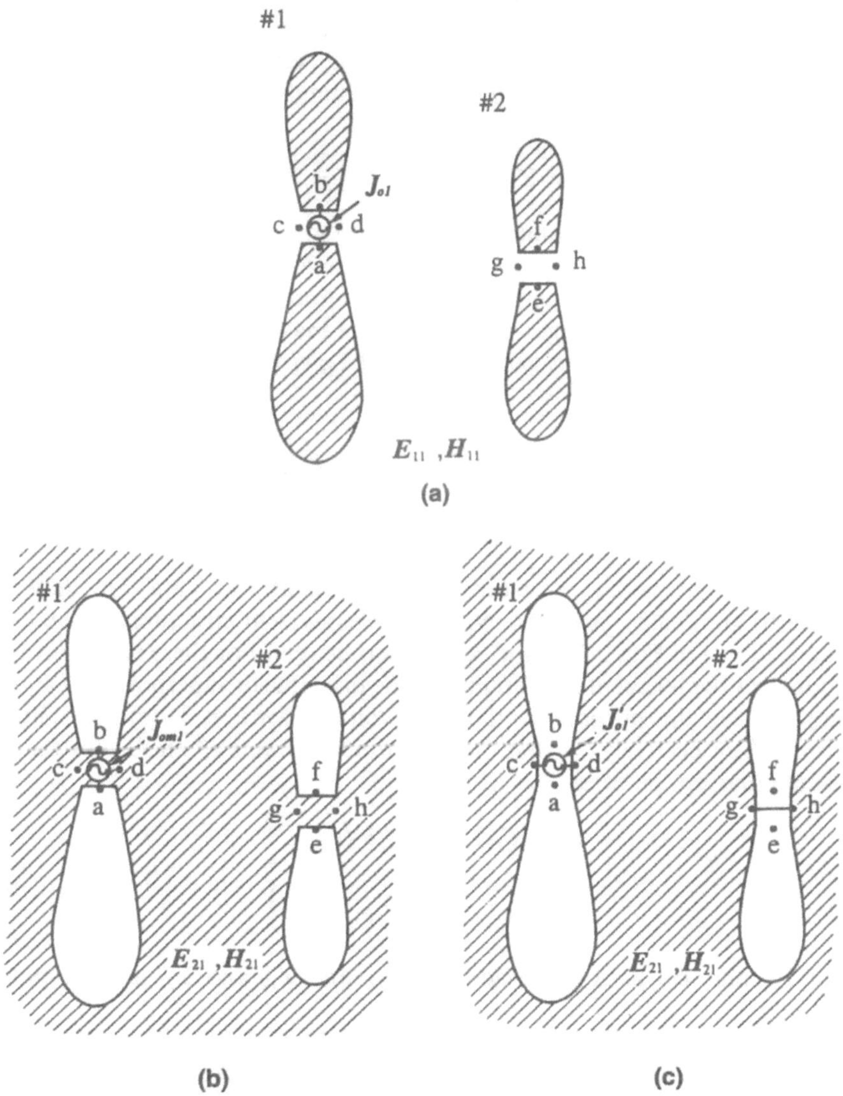

Fig. 3.2 Excitation of mutually complementary two-element planar structures (Case 1).

In Fig. 3.2(a), antenna 1 of the two co-planar plate antennas is excited by electric source current I_1, and antenna 2 is opened. Then, the induced electromagnetic fields, $\boldsymbol{E}_{11}$, $\boldsymbol{H}_{11}$, for this case are proportional to I_1, and hence the terminal voltages, V_{11}, V_{21}, of antennas 1 and 2 are also proportional to I_1. Accordingly, the terminal voltages can be written as

$$V_{11} = Z_{11}I_1 \quad \text{and} \quad V_{21} = Z_{21}I_1 \tag{3.4}$$

where Z_{11} is the self-impedance of antenna 1, and Z_{21} is the mutual impedance between the two antennas.

Next, let us consider another case, as shown in Fig. 3.3(a), where antenna 1 is opened and antenna 2 is excited by the electric source current I_2. Then, the induced electromagnetic fields, $\boldsymbol{E}_{12}$, $\boldsymbol{H}_{12}$, and the terminal voltages V_{12}, V_{22}, of the two antennas are proportional to I_2. Accordingly, the terminal voltages can be written as

$$V_{12} = Z_{12}I_2 \quad \text{and} \quad V_{22} = Z_{22}I_2 \tag{3.5}$$

where Z_{12} and Z_{22} are the mutual impendance and the self-impedance.

A general case, where both antennas are excited at the same time, can be treated by applying the principle of superposition for electromagnetic fields. Let the total terminal voltages of the two antennas be V_1 and V_2, when both are excited by electric source currents I_1 and I_2, respectively. Then, by taking the sum of (3.4) and (3.5) we obtain

$$\left.\begin{aligned} V_1 &= Z_{11}I_1 + Z_{12}I_2 \\ V_2 &= Z_{21}I_1 + Z_{22}I_2 \end{aligned}\right\} \tag{3.6}$$

Now, the complementary structures for Figs 3.2(a) and 3.3(a) are shown in Figs 3.2(b) and 3.3(b), respectively, and the equivalent structures of the latter ones are shown in Figs 3.2(c) and 3.3(c), respectively. For these two equivalent structures of two-element hole antennas, the principle of superposition can also be applied. For this purpose, let the input voltages for Figs 3.2(c) and 3.3(c) be V_1' and V_2', respectively, so that the total input currents I_1' and I_2' for the two hole antennas are given by

$$\left.\begin{aligned} I_1' &= Y_{11}'V_1' + Y_{12}'V_2' \\ I_2' &= Y_{21}'V_1' + Y_{22}'V_2' \end{aligned}\right\} \tag{3.7}$$

where Y_{11}' and Y_{22}' are the self-admittances, and Y_{12}' and Y_{21}' are the mutual admittances.

Now, the electric source currents in Figs 3.2(a) and 3.3(b) can be regarded as symmetrical, and the magnetic source currents in Figs 3.2(b) and 3.3(b) as antisymmetrical with respect to the planar sheets. Accordingly, the duality relationships corresponding to (3.1) can be expressed as:

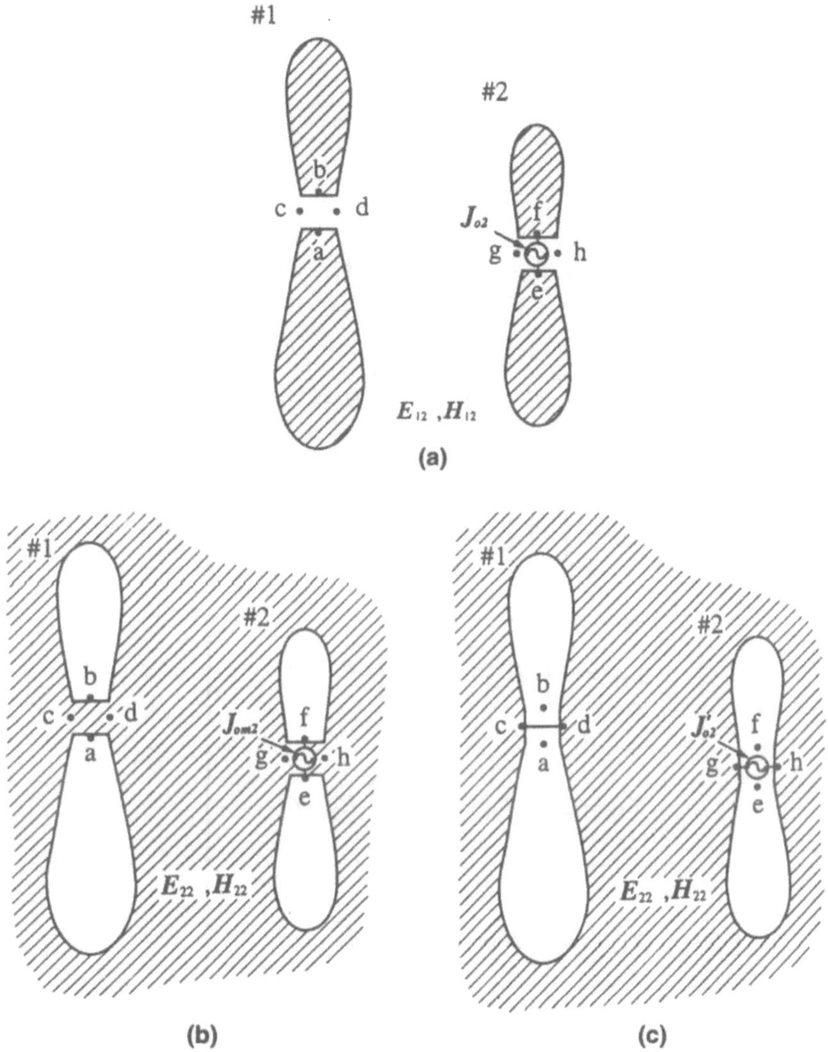

Fig. 3.3 Excitation of mutually complementary two-element planar structures (Case 2).

$$\left.\begin{aligned} \boldsymbol{E}_{21} &= \mp K_1 \boldsymbol{H}_{11}, \qquad & \boldsymbol{H}_{21} &= \pm K_1 \gamma \boldsymbol{E}_{11} \\ \boldsymbol{E}_{22} &= \mp K_2 \boldsymbol{H}_{12}, \qquad & \boldsymbol{H}_{22} &= \pm K_2 \gamma \boldsymbol{E}_{12} \end{aligned}\right\} \tag{3.8}$$

where K_1, K_2 are constants which depend on the ratios of the electric source currents for the mutually complementary structures in Figs 2.2 and 2.3, respectively. Hence, the impedances in (3.6) and the admittances in (3.7) can be expressed by the corresponding electromagnetic fields in a similar way to that of (3.2), as follows:

$$\left.\begin{aligned}
Z_{11} &= \int_b^a \boldsymbol{E}_{11} \cdot \mathrm{d}\boldsymbol{l} \bigg/ 2\int_c^d \boldsymbol{H}_{11} \cdot \mathrm{d}\boldsymbol{l} \\
Z_{21} &= \int_f^e \boldsymbol{E}_{11} \cdot \mathrm{d}\boldsymbol{l} \bigg/ 2\int_c^d \boldsymbol{H}_{11} \cdot \mathrm{d}\boldsymbol{l} \\
Z_{12} &= \int_b^a \boldsymbol{E}_{12} \cdot \mathrm{d}\boldsymbol{l} \bigg/ 2\int_g^h \boldsymbol{H}_{12} \cdot \mathrm{d}\boldsymbol{l} \\
Z_{22} &= \int_f^e \boldsymbol{E}_{12} \cdot \mathrm{d}\boldsymbol{l} \bigg/ 2\int_g^h \boldsymbol{H}_{12} \cdot \mathrm{d}\boldsymbol{l}
\end{aligned}\right\} \tag{3.9}$$

$$\left.\begin{aligned}
Y'_{11} &= 2\int_b^a \boldsymbol{H}_{21} \cdot \mathrm{d}\boldsymbol{l} \bigg/ \int_d^c \boldsymbol{E}_{21} \cdot \mathrm{d}\boldsymbol{l} \\
Y'_{21} &= 2\int_f^e \boldsymbol{H}_{21} \cdot \mathrm{d}\boldsymbol{l} \bigg/ \int_d^c \boldsymbol{E}_{21} \cdot \mathrm{d}\boldsymbol{l} \\
Y'_{12} &= 2\int_b^a \boldsymbol{H}_{22} \cdot \mathrm{d}\boldsymbol{l} \bigg/ \int_h^g \boldsymbol{E}_{22} \cdot \mathrm{d}\boldsymbol{l} \\
Y'_{22} &= 2\int_f^e \boldsymbol{H}_{22} \cdot \mathrm{d}\boldsymbol{l} \bigg/ \int_h^g \boldsymbol{E}_{22} \cdot \mathrm{d}\boldsymbol{l}
\end{aligned}\right\} \tag{3.10}$$

By introducing relationships (3.8) into the integrands of (3.9) and (3.10), we find

$$\left.\begin{aligned}
Y'_{11} &= 4\gamma Z_{11}, \qquad Y'_{12} = 4\gamma Z_{12} \\
Y'_{21} &= 4\gamma Z_{21}, \qquad Y'_{22} = 4\gamma Z_{22}
\end{aligned}\right\} \tag{3.11}$$

These relationships can be expressed in matrix form as

$$[Y'] = 4\gamma[Z] \tag{3.12}$$

or

$$[Z'] = [Y']^{-1} = [Z]^{-1}/4\gamma = [Z]^{-1}(Z_0/2)^2 \quad (3.13)$$

In general, n-element mutually complementary, co-planar structures can be treated similarly, and their impedance relationships are also given by expressions (3.12) or (3.13).

3.3 Examples of complementary planar structures

Some typical examples of mutually complementary planar structures where impedance relationship (3.3) can be applied, will be shown in this section, though numerous other structures are conceivable.

Well-known examples of complementary structures are the conducting sheet strip antenna and the slot antenna as shown in Figs 3.4(a) and (b). When the width, W, of the strip and the slot is sufficiently small compared with the wavelength, a long strip antenna is equivalent to a wire antenna with a circular cross-section. The equivalent radius, ρ_e, of such an antenna is given [1.18] by

$$\rho_e = W/4 \quad (3.14)$$

Accordingly, various characteristics of a slot antenna with a narrow width can easily be obtained from those of a corresponding cylindrical antenna which has a circular cross-section with the equivalent radius mentioned above.

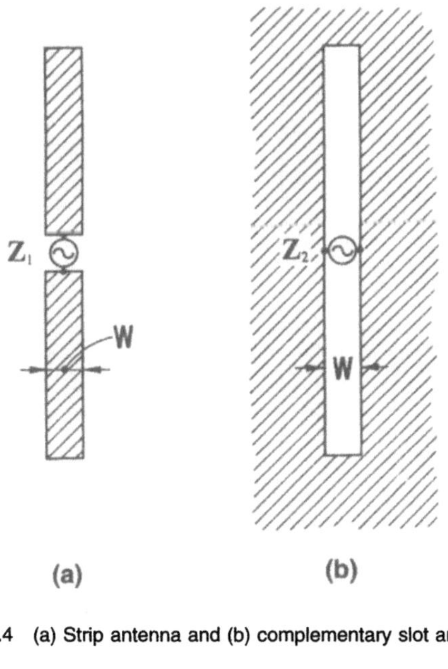

Fig. 3.4 (a) Strip antenna and (b) complementary slot antenna.

Other examples of mutually complementary antennas are shown in Fig. 3.5. Though the slot of Fig. 3.4(b) is made in an infinitely extended conducting planar sheet, that of Fig. 3.5(b) is made in a semi-infinite conducting plane. However, the following property can be deduced from Fig. 3.5: the effects of truncation of the conducting sheet for the slot antenna can be determined from the effects of the introduction of the semi-infinite conducting sheet for the strip antenna, as shown in Fig. 3.5(a). This is so because the corresponding electromagnetic fields for the two cases are always in the relationship given in (3.1).

Various characteristics of a notch antenna, as shown in Fig. 3.6(b), can be obtained by analysing the unipole (or monopole) antenna located at the edge of a semi-infinite plane of conducting sheet, as shown in Fig. 3.6(a) [3.5].

The structures shown in Figs 3.7(a) and (b) can be considered as types of transmission lines, if the transverse dimensions of the strips are sufficiently small compared with the wavelength. In such cases, their characteristic impedances, Z_1 and Z_2, become pure resistances, see relationship (3.3), and no radiation takes place from these structures.

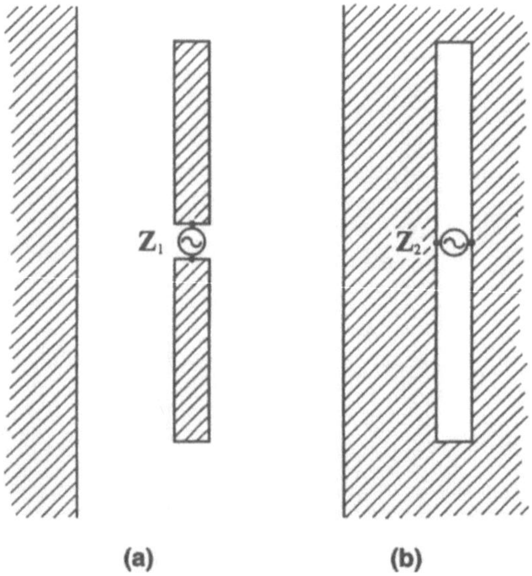

Fig. 3.5 (a) Strip antenna with semi-infinite plane of perfect conductor and (b) its complementary structure (slot antenna on semi-infinite plane of perfect conductor).

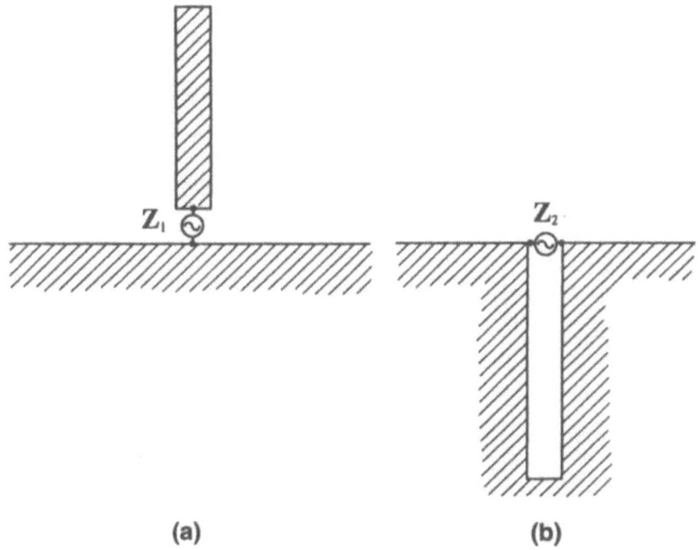

Fig. 3.6 (a) Unipole antenna and (b) complementary notch antenna.

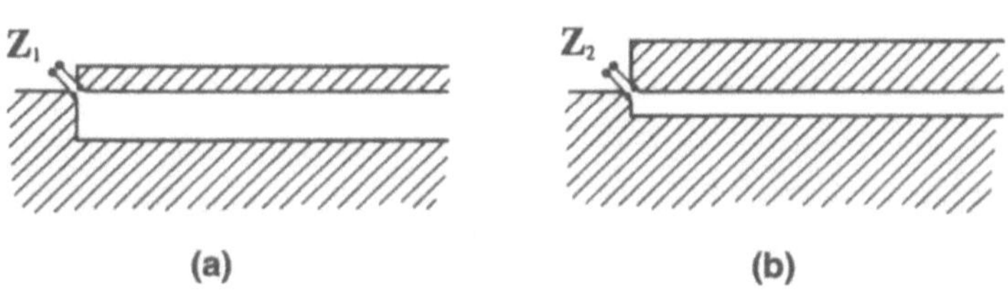

Fig. 3.7 Mutually complementary strip lines located along the edges of a semi-infinite planar sheet of perfect conductor.

4 • ORIGINATION OF SELF-COMPLEMENTARY PLANAR STRUCTURES AND DISCOVERY OF THEIR CONSTANT-IMPEDANCE PROPERTY

4.1 Origination of self-complementary planar structures

According to the result obtained in Chapter 3, section 3.1, the input impedances, Z_1 and Z_2, for mutually complementary planar structures are given by relationship (3.3), that is

$$Z_1 Z_2 = (Z_0/2)^2 \tag{4.1}$$

where Z_0 is the intrinsic impedance of the medium, which is approximately equal to 120π [Ω] in free space. This relationship was derived by the author in 1948 [1.1–1.3]. As mentioned in section 3.1, however, the same expression had already been obtained by several other investigators, after making various assumptions, as the relationship between the input impedances for a slot antenna and its complementary wire antenna [3.1–3.4]. However, expression (4.1) in the present theory is derived without any assumptions being made, and it is always exact for any shape of mutually complementary structure. Therefore, expression (4.1) is an innovative and generalized relationship for a pair of arbitrarily shaped complementary planar structures. Nevertheless, the originality or novelty of the author's theory was not appreciated when it first appeared.

For that reason, the author's main aim was to describe some definite examples which exhibited the distinctive feature of this relationship (4.1), and intensive considerations and studies were hence continued over a fairly long period. After repeated attempts to this end, an entirely new concept became apparent to the author. This was the self-complementary planar structure, where the shape of the original sheet structure is identical to that of its complementary planar structure. In other words, the self-complementary structure is a structure that is complementary to the original structure itself.

The simplest structures that were shown in the original papers to illustrate this principle are reproduced in Fig. 4.1 [1.1—1.3].

The points indicated by "a" and "b" in the figures are feed terminals. The structure in Fig. 4.1(a) is rotationally symmetric, where four contour lines of two conducting sheets are formed from one arbitrarily curved generating line by rotating it through 90° steps. In contrast, the structure in Fig. 4.1(b) is axially symmetric, where one of two contour lines for both of two conducting sheets are on the axis of symmetry, while the other contour lines are formed from axially symmetric curved lines with an arbitrary shape.

The examples shown are two fundamental types of self-complementary planar structure. In these structures, the contour lines can be deformed in any way, even to the extent where one edge line is composed of several separated lines, rather than a single continuous curved line. Examples of such structures derived from sophisticated edge lines are shown in Figs 4.2(a) and (b).

4.2 Constant-impedance property of self-complementary planar structures

The input impedances of the self-complementary planar structures derived in the preceding section will be investigated in this section. As a natural consequence of the self-complementary configuration of the structure, the input impedance Z_1 of the original structure is identical to that of its complementary structure Z_2, and they have a common value Z. That is

$$Z_1 = Z_2 = Z \tag{4.2}$$

Introducing this relationship into (4.1), we obtain

$$Z = Z_0/2 \tag{4.3}$$

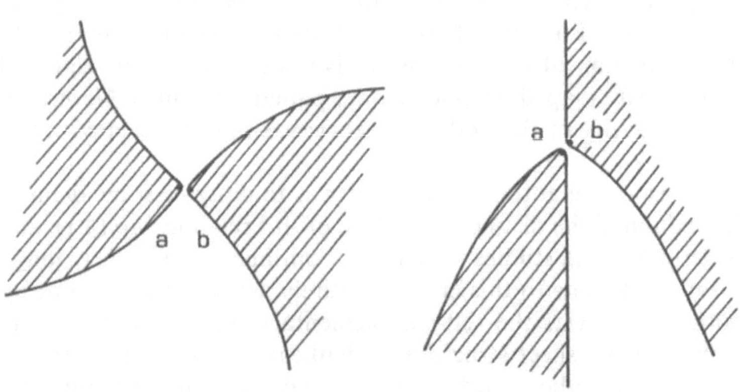

(a) (b)

Fig. 4.1 Simplified shapes of self-complementary structures to illustrate the principle.

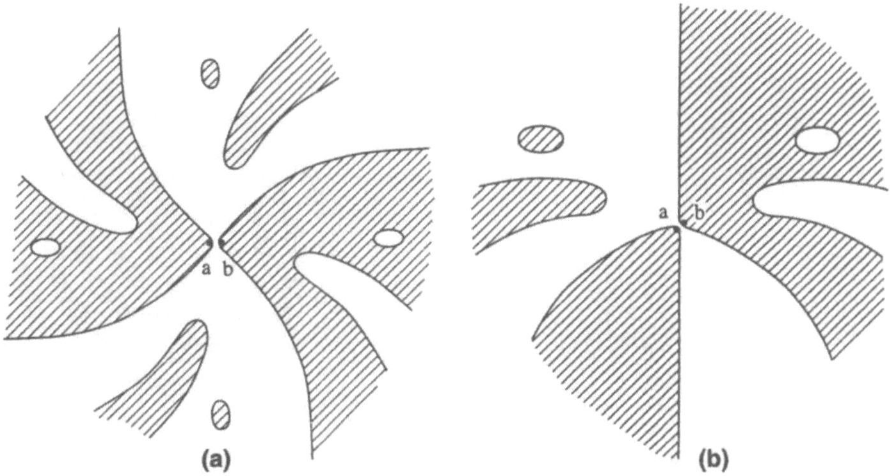

Fig. 4.2 Sophisticated shapes of self-complementary structures to illustrate the principle.

which in free space reduces to

$$Z \simeq 60\pi \simeq 188 \qquad [\Omega] \tag{4.4}$$

Expressions (4.3) and (4.4) mean that the input impedance of the self-complementary planar structure is always constant whatever the source frequency and shape of the structure. As this remarkable constant-impedance property was discovered by the present author, expression (4.3) is called "Mushiake's relationship" [1.11]. Fortunately, the input impedance given by (4.3) or (4.4) is adequate for an antenna in practical use. Therefore, the constant-impedance property of self-complementary structures provides an important and effective technological guidance principle for developmental investigations of extremely broad-band practical antennas.

The constant-impedance property described above is the distinctive characteristic of a particular planar structure whose geometry is identical with that of its complementary planar structure. However, such a property is not restricted only to structures made from a single planar sheet. In practice, the principle can be extended from the basic structure to various other cases having rather more complicated structure, as will be explained in the following chapters. For this reason, the fundamental *Principle of self-complementarity* discovered by the author can be extensively applied in a broad sense.

4.3 Examples of self-complementary structures

Some examples of self-complementary antennas corresponding to the two fundamental types are shown in Figs 4.3 and 4.4.

The structures in Fig. 4.3 are rotationally symmetric, and they work as balanced-type constant-impedance antennas. In contrast, the structure in

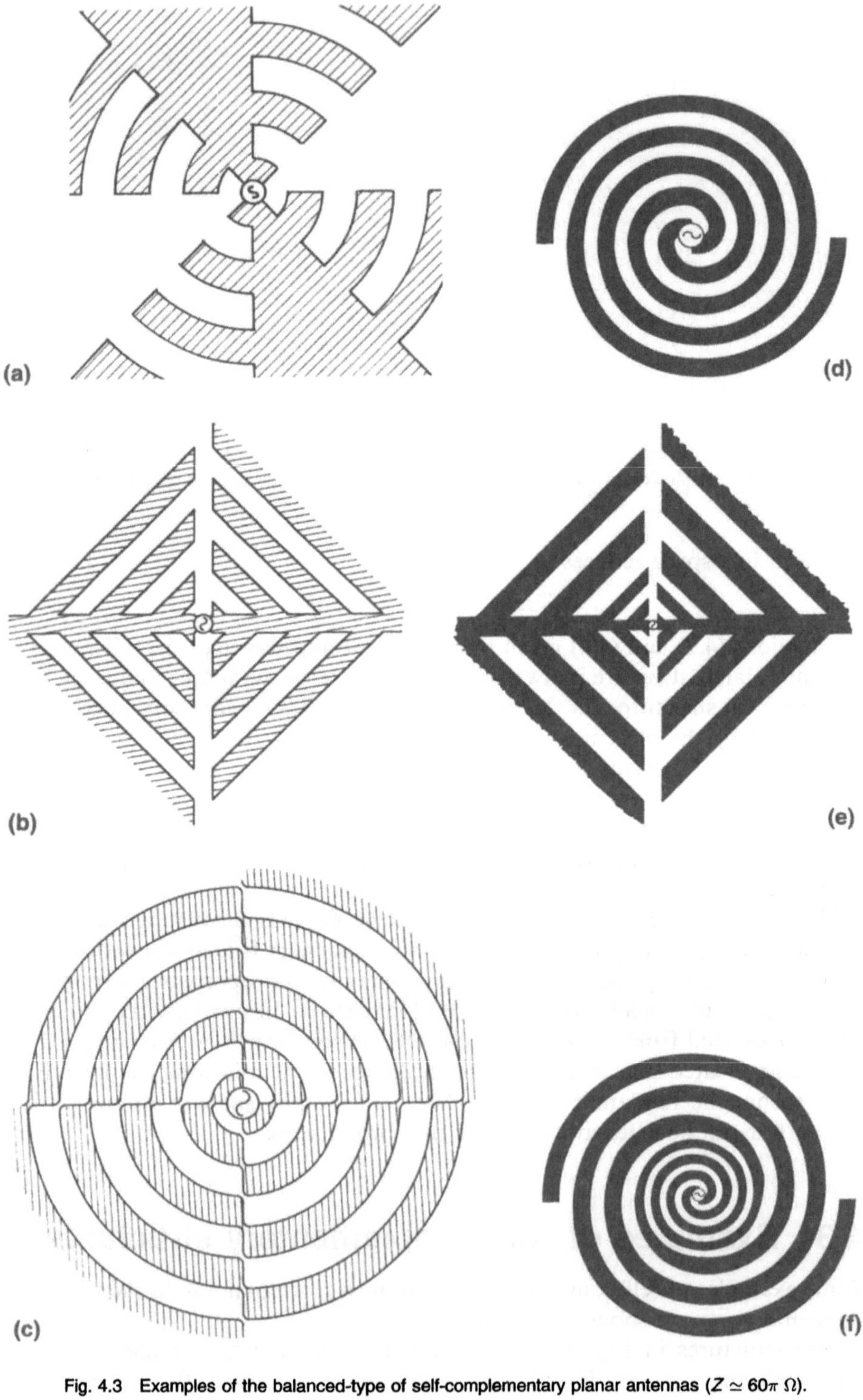

Fig. 4.3 Examples of the balanced-type of self-complementary planar antennas ($Z \simeq 60\pi\ \Omega$).

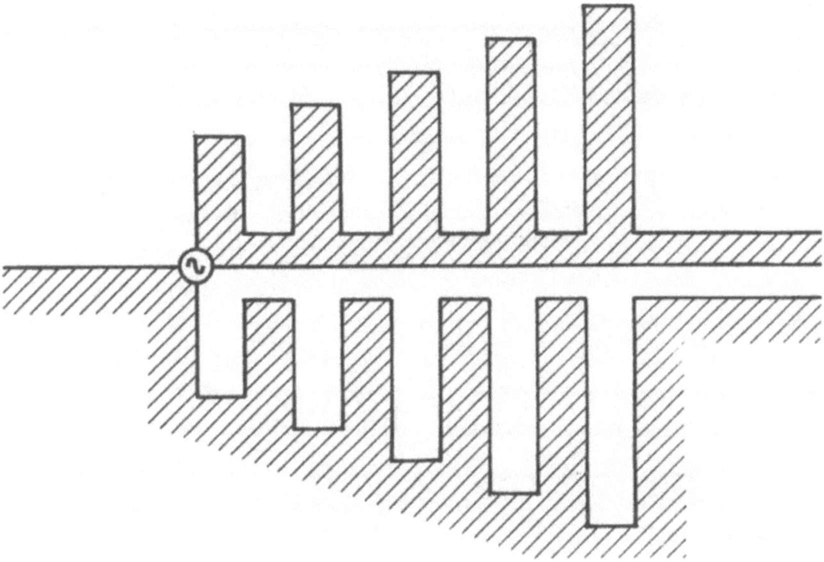

Fig. 4.4 An example of an unbalanced-type of self-complementary planar antenna ($Z \simeq 60\pi\ \Omega$).

Fig. 4.4 is axially symmetric, and it works as an unbalanced-type constant-impedance antenna. In these structures, Figs 4.3(a) – (d) and Fig. 4.4 are in equally spaced forms, but Figs 4.3(e) and (f) are in compound-spaced forms. Furthermore, proportionally spaced forms, or log-periodic forms, and various other forms of constant-impedance antennas, could be easily conceived via this principle.

Also, for the structure shown in Fig. 4.5(a), if the transverse dimensions of the strip and the gap are sufficiently small compared with the wavelength, then this structure works as a non-radiative transmission line with characteristic impedance $Z_0/2$. Therefore, an infinitely long portion of the strip in Fig. 4.5(a) can be substituted by a load impedance $Z_0/2$, as shown in Fig. 4.5(b), without disturbing any electromagnetic fields. Similarly, an infinitely long strip in Fig. 4.4 can also be substituted by a load impedance $Z_0/2$ without disturbing any characteristics of that antenna. For interest, Figs 4.5(c) and (d) show series and parallel connections of such self-complementary transmission lines.

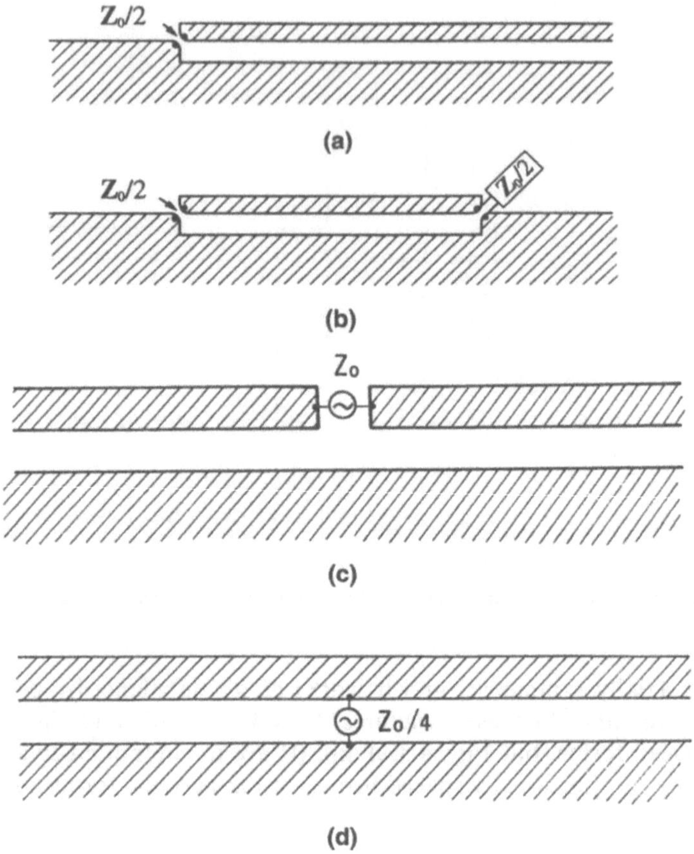

Fig. 4.5 Self-complementary transmission line (constituted along the edge of a semi-infinite planar sheet of perfect conductor): (a) without termination, (b) terminated by $Z_0/2$ (both $Z \simeq 60\pi\ \Omega$), (c) series connection ($Z \simeq 120\pi\ \Omega$), (d) parallel connection ($Z \simeq 30\pi\ \Omega$).

5 • MULTI-TERMINAL SELF-COMPLEMENTARY PLANAR STRUCTURES

5.1 Rotationally symmetric multi-terminal self-complementary planar structures

A simple and typical example of a rotationally symmetric multi-terminal self-complementary planar structure is shown in Fig. 5.1, where all of the eight contour lines of the four conducting planar sheets are generated with a single bent line by rotating it by 45° steps.

Now, let us assume that the four terminals are excited by symmetrical four-phase electric sources in star connection, as shown in Fig. 5.2(a).

By examining the directions of the magnetic field vectors at these feed points, it is found that the electric sources are equivalent to the magnetic current sources in ring connection, respectively on the front side and the reverse side

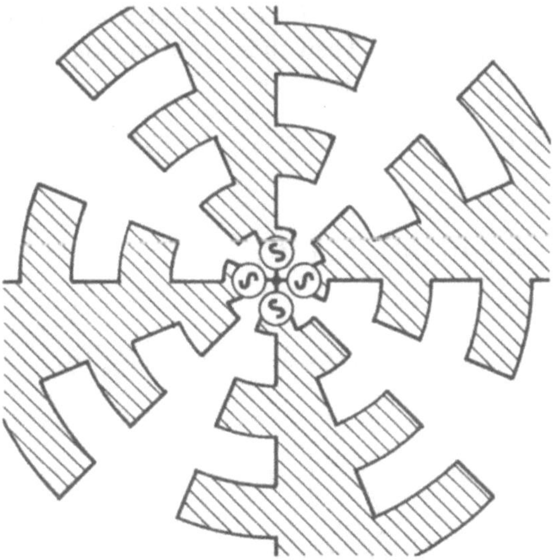

Fig. 5.1 An example of a four-terminal self-complementary planar turnstile antenna ($Z \simeq 30\sqrt{2}\pi\ \Omega$).

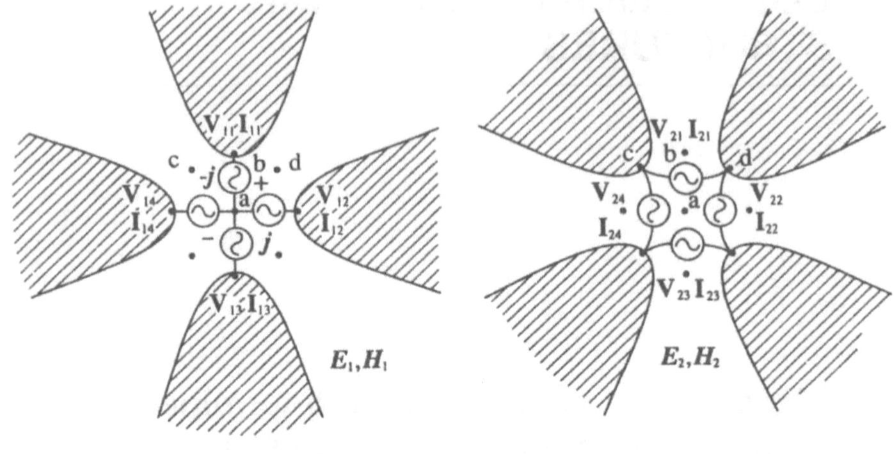

(a) (b)

Fig. 5.2 Four-terminal structure excited by four-phase electric sources: (a) star connection, (b) ring connection.

of the planar structure. Therefore, the source currents in its complementary structure are in ring connection as shown in Fig. 5.2(b), though the shapes of the original and the complementary planar structures are identical. Incidentally, it should be noted here that the excitation in Fig. 5.2(b) corresponds to that in Fig. 3.1(c) in the case of a two-terminal structure, or to that in Fig. 3.2(c) in the case of a two-element structure.

To deal with the structure in Fig. 5.2 that is similar to that of a two-terminal in Fig. 3.1, let the input impedances for each electric source in Figs 5.2(a) and (b) be Z_1 and Z_2, respectively. Then these two impedances are given by the same expressions as (3.2) [1.4,1.6], that is

$$\left.\begin{aligned} Z_1 &= V_{11}/I_{11} = V_{12}/I_{12} = V_{13}/I_{13} = V_{14}/I_{14} \\ &= \int_b^a \boldsymbol{E}_1 \cdot \mathrm{d}\boldsymbol{l} \Big/ 2\int_c^d \boldsymbol{H}_1 \cdot \mathrm{d}\boldsymbol{l} \\ Z_2 &= V_{21}/I_{21} = V_{22}/I_{22} = V_{23}/I_{23} = V_{24}/I_{24} \\ &= \int_d^c \boldsymbol{E}_2 \cdot \mathrm{d}\boldsymbol{l} \Big/ 2\int_b^a \boldsymbol{H}_2 \cdot \mathrm{d}\boldsymbol{l} \end{aligned}\right\} \tag{5.1}$$

By taking the product of the expressions for Z_1 and Z_2, and introducing relationships (1.25) into the integrands, we find

$$Z_1 Z_2 = 1/4\gamma = (Z_0/2)^2 \tag{5.2}$$

This is exactly the same expression as (3.2) for two-terminal structures.

However, the input impedances for electric sources in star connection are different from those in ring connection, when they are connected to the same multi-terminal loads. Accordingly, this difference must be taken into account in the present treatment. For this purpose, the impedances for the electric sources in two types of connection as shown in Figs 5.3(a) and (b), will be analysed.

As indicated in the figures, let V_s, I_s and V_r, I_r be the voltages and the currents of the four-phase electric sources respectively in star connection and ring connection. For these two cases, identical four-terminal symmetrical loads are assumed. Then, we find

$$\left.\begin{aligned} V_r &= jV_s - V_s \\ I_s &= -jI_r - I_r \end{aligned}\right\} \tag{5.3}$$

and hence, the impedance Z_s for each electric source in star connection, and the impedance Z_r for each electric source in ring connection, are related by

$$Z_r = V_r/I_r = 2V_s/I_s = 2Z_s \tag{5.4}$$

By applying this result to the two geometrically identical structures in Fig. 5.2, we find

$$Z_2 = Z_r = 2Z_s = 2Z_1 \tag{5.5}$$

This means that the two structures in Fig. 5.2 are geometrically self-complementary but the impedances are not identical, owing to the difference in their configurations as a whole. The introduction of this relationship into (5.2) for the elimination of Z_2 leads to

$$Z_1 = Z_0/2\sqrt{2} \simeq 30\sqrt{2}\,\pi \quad [\Omega] \tag{5.6}$$

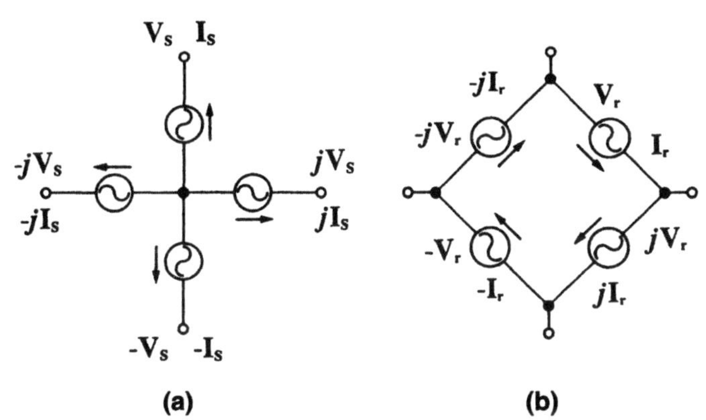

Fig. 5.3 Four-phase electric sources and their equivalent transformation: (a) star connection, (b) ring connection.

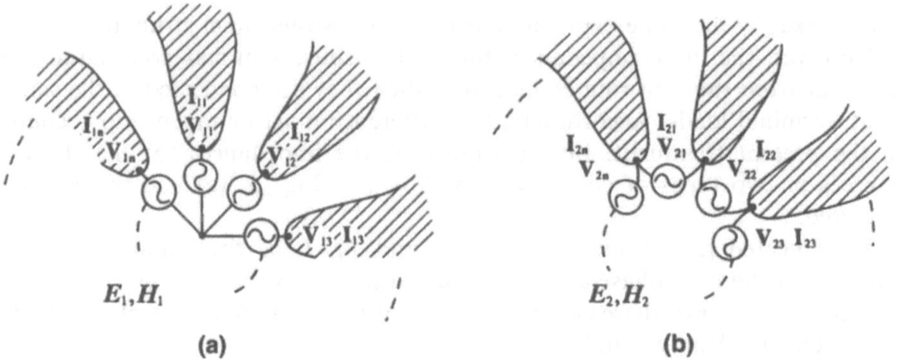

Fig. 5.4 *n*-terminal structure excited by *n*-phase electric sources: (a) star connection, (b) ring connection.

Next, the general case shown in Figs 5.4(a) and (b), where a rotationally symmetric n-terminal self-complementary planar structure is excited by symmetrical n-phase electric sources in star and ring connections with mth-order rotation, is considered. Here, the same expressions as (5.1) and (5.2) also hold for these structures. However, the relationship between the two impedances in the star connection and in the ring connection is not the same as (5.5).

Referring to Figs 5.5(a) and (b), we find that relationships (5.3) must be replaced by

$$\left.\begin{aligned} V_r &= \delta V_s - V_s \\ I_s &= \delta^{-1} I_r - I_r \end{aligned}\right\} \tag{5.7}$$

where

$$\delta = \exp(\mathrm{j}2m\pi/n) \tag{5.8}$$

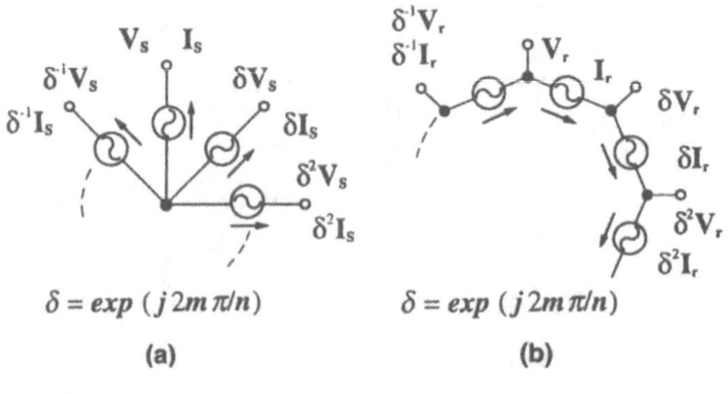

Fig. 5.5 *n*-phase electric sources with *m*th-order rotation and their equivalent transformation: (a) star connection, (b) ring connection.

Hence, we have

$$Z_r = V_r/I_r = 4\{\sin(m\pi/n)\}^2 V_s/I_s$$
$$= \{2\sin(m\pi/n)\}^2 Z_s \tag{5.9}$$

Accordingly, the impedances Z_1 and Z_2 for the two geometrically identical structures in Fig. 5.4 are related by

$$Z_2 = \{2\sin(m\pi/n)\}^2 Z_1 \tag{5.10}$$

Consequently, the input impedance Z_1 for star connection in Fig. 5.4(a) is given by

$$Z_1 = Z_0/4\sin(m\pi/n) \tag{5.11}$$

Expressions (5.6) and (5.11) mean that the input impedances for rotationally symmetric multi-terminal self-complementary planar structures are always constant, independently of the source frequency and of the shape of the planar structures.

5.2 Single-phase excitations for rotationally symmetric multi-terminal self-complementary structures

The input impedances for various single-phase excitations of rotationally symmetric multi-terminal self-complementary antennas will be discussed in detail in this section.

Let Z_{1mn} be the input impedance of rotationally symmetric n-terminal self-complementary planar structures excited by n-phase electric sources in star connection with mth-order rotation. Then, Z_{1mn} is given by (5.11) as

$$Z_{1mn} = Z_0/4\sin(m\pi/n) \tag{5.12}$$

For the simplest case of $n = 3$, (5.12) reduces to

$$Z_{113} = Z_{123} = Z_0/2\sqrt{3}$$
$$\simeq 20\sqrt{3}\,\pi \quad [\Omega] \tag{5.13}$$

The value of the impedance given by (5.13) is for each phase of the three-phase electric source in star connection. However, if the three terminals for this structure are rearranged as shown in Fig. 5.6, where in (a) two terminals are short-circuited, and in (b) one terminal is floating, then the three terminals can be excited by a single-phase electric source, and the values of the input impedance must be different from the original one.

In order to obtain the input impedances for such single-phase excitations, two cases of rearranged connections as shown in Figs 5.6(a) and (b) are con-

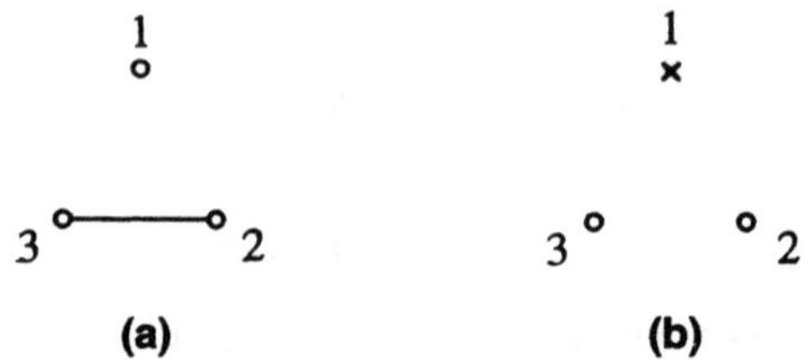

Fig. 5.6 Single-phase connections of three terminals.

sidered. In the case of three-phase electric sources, all possible modes are for $m = 1$ and $m = 2$, and the mode voltages v_{mn} can be assumed for these two modes as follows:

$$\left.\begin{array}{ll} v_{11} = A_1, & v_{21} = A_2 \\ v_{12} = A_1 \exp(\mathrm{j}2\pi/3), & v_{22} = A_2 \exp(-\mathrm{j}2\pi/3) \\ v_{13} = A_1 \exp(-\mathrm{j}2\pi/3), & v_{23} = A_2 \exp(\mathrm{j}2\pi/3) \end{array}\right\} \tag{5.14}$$

The condition for Fig. 5.6(a) is given by

$$v_{12} + v_{22} = v_{13} + v_{23}$$

which leads to

$$A_1 = A_2 \tag{5.15}$$

Accordingly, the total voltage V_t between terminals "1" and "2" is given by

$$\begin{aligned} V_t &= 2A_1 - A_1\{\exp(\mathrm{j}2\pi/3) + \exp(-\mathrm{j}2\pi/3)\} \\ &= 4A_1 \sin^2(\pi/3) = 3A_1 \end{aligned} \tag{5.16}$$

On the other hand, the total current I_t for terminal "1" is given by

$$I_t = (v_{11}/Z_{113}) + (v_{21}/Z_{123}) = 2A_1/Z_{113} \tag{5.17}$$

By taking the ratio of V_t and I_t, the input impedance Z_t for the connection in Fig. 5.6(a) can be determined as

$$Z_t = 3Z_{113}/2 = \sqrt{3}\, Z_0/4 \simeq 30\sqrt{3}\,\pi \quad [\Omega] \tag{5.18}$$

Similar consideration of Fig. 5.6(b) leads to

$$A_1 = -A_2 \tag{5.19}$$

Accordingly, the total voltage V_t between terminals "1" and "2", the input current I_t, and the input impedance Z_t for the connection in Fig. 5.6(b), are respectively determined as

$$\left.\begin{aligned} V_t &= \mathrm{j}4A_1 \sin(2\pi/3) \\ I_t &= \mathrm{j}2A_1 \sin(2\pi/3)/Z_{113} \\ Z_t &= 2Z_{113}Z_0/\sqrt{3} \simeq 40\sqrt{3}\,\pi \quad [\Omega] \end{aligned}\right\} \tag{5.20}$$

Next, let us deal with the four-terminal symmetrical structure. In this case, three modes of four-phase electric sources must be taken into account, corresponding to the two modes in (5.14) for three-phase sources. The four-phase voltages in star connection, and the impedances for three modes are listed in Table 5.1.

Table 5.1 Four-phase voltages in star connection and the input impedances for three modes

Terminal	Mode		
	$m = 1$	$m = 2$	$m = 3$
1	A_1	A_2	A_3
2	$\mathrm{j}A_1$	$-A_2$	$\mathrm{j}A_3$
3	$-A_1$	A_2	A_3
4	$-\mathrm{j}A_1$	$-A_2$	$\mathrm{j}A_3$
Impedance (star conn.)	$Z_0/2\sqrt{2}$	$Z_0/4$	$Z_0/2\sqrt{2}$

After some straightforward calculations, the total input impedances of the four-terminal structure for various connections of single-phase excitation can be determined. The results of the calculations are listed in Table 5.2.

Further calculations for larger values of n could be performed. Some detailed examples were shown by G.A. Deschamps [1.21].

5.3 Axially symmetric multi-terminal self-complementary planar structures

An example of a three-terminal geometrically self-complementary planar structure as shown in Fig. 5.7(a) is considered.

In this structure, an electric source V_{11} is connected between terminals "a" and "b", but terminals "b" and "d" are terminated by a load impedance Z_{l1} instead of an electric source. Its complementary structure is shown in Fig. 5.7(b), where an electric source V_{21} and a load impedance Z_{l2} are connected between the terminals as illustrated. In these two structures, their shapes are geometrically self-complementary. However, when the configurations as a whole, including the conditions for the feed points, are taken into account, there is no assurance of the self-complementary nature. In addition, even the mutually complementary nature is not guaranteed in general.

Table 5.2 Total input impedances of four-terminal self-complementary antennas for various single-phase connections

Connection	Total input impedance	
	Expression	Value in free space (Ω)
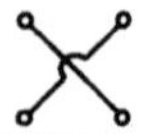	$Z_0/4$	$30\pi \simeq 94.2$
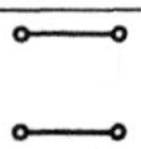	$Z_0/2\sqrt{2}$	$30\sqrt{2}\pi \simeq 133.2$
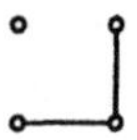	$Z_0/(1+\sqrt{2})$	$\dfrac{120\pi}{1+\sqrt{2}} \simeq 156.1$
	$\dfrac{Z_0(1+\sqrt{2})}{4\sqrt{2}}$	$(2+\sqrt{2})15\pi \simeq 160.8$
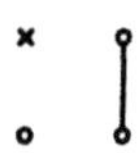	$Z_0/2$	$60\pi \simeq 188.4$
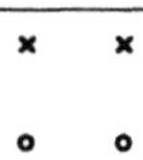	$Z_0(1+\sqrt{2})/4$	$(1+\sqrt{2})30\pi \simeq 227.4$
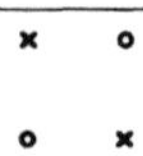	$Z_0/\sqrt{2}$	$60\sqrt{2}\pi \simeq 266.4$

o: feeding terminals in two groups.
×: floating terminals.

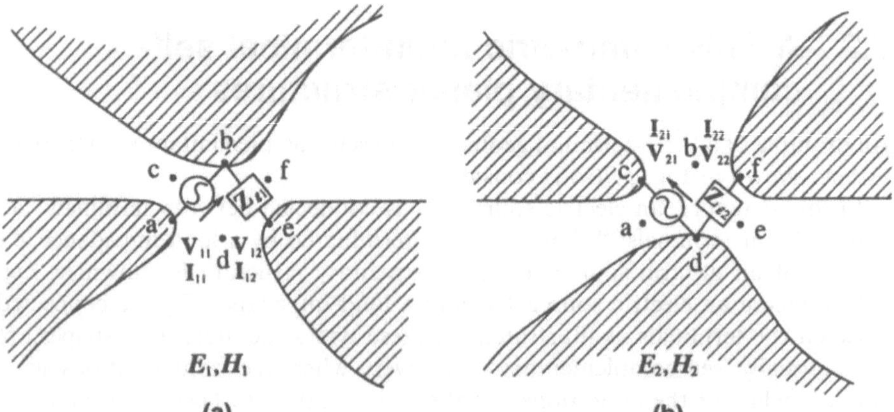

(a) (b)

Fig. 5.7 Axially symmetric three-terminal loaded self-complementary planar structure.

However, let us assume here that these two structures are mutually complementary. Then, the input impedances Z_1, Z_2 and the load impedances Z_{l1}, Z_{l2} at their terminals can be expressed as follows:

$$\left.\begin{aligned} Z_1 &= V_{11}/I_{11} = \int_b^a \boldsymbol{E}_1 \cdot \mathrm{d}\boldsymbol{l} \Big/ 2\int_c^d \boldsymbol{H}_1 \cdot \mathrm{d}\boldsymbol{l} \\ Z_2 &= V_{21}/I_{21} = \int_d^c \boldsymbol{E}_2 \cdot \mathrm{d}\boldsymbol{l} \Big/ 2\int_b^a \boldsymbol{H}_2 \cdot \mathrm{d}\boldsymbol{l} \end{aligned}\right\} \tag{5.21}$$

$$\left.\begin{aligned} Z_{l1} &= V_{12}/I_{12} = \int_b^e \boldsymbol{E}_1 \cdot \mathrm{d}\boldsymbol{l} \Big/ 2\int_f^d \boldsymbol{H}_1 \cdot \mathrm{d}\boldsymbol{l} \\ Z_{l2} &= V_{22}/I_{22} = \int_d^f \boldsymbol{E}_2 \cdot \mathrm{d}\boldsymbol{l} \Big/ 2\int_b^e \boldsymbol{H}_2 \cdot \mathrm{d}\boldsymbol{l} \end{aligned}\right\} \tag{5.22}$$

Introducing relationships (3.1) into the integrands of these expressions, we find

$$Z_1 Z_2 = (Z_0/2)^2 \tag{5.23}$$

$$Z_{l1} Z_{l2} = (Z_0/2)^2 \tag{5.24}$$

Expression (5.24) is a condition for the load impedances to make these structures mutually complementary, while relationship (5.23) is derived as a consequence of the mutually complementary nature. Furthermore, if load impedances Z_{l1} and Z_{l2} are identical, then the two structures become identical and they are self-complementary. Accordingly, from condition (5.24) we find

$$Z_{l1} = Z_{l2} = Z_0/2 \tag{5.25}$$

Under this condition, relationship (5.23) yields

$$Z_1 = Z_2 = Z_0/2 \tag{5.26}$$

However, if condition (5.24) is not satisfied, then the two structures are not mutually complementary. Therefore, no relationship such as (5.23) or (5.26) can be obtained between the two structures in Fig. 5.7 unless (5.24) is satisfied.

Also, if load impedances Z_{l1}, Z_{l2} in Fig. 5.7 are replaced by electric sources, then the relative directions of the electric sources in the respective structures cannot be identical for these two structures. Accordingly, their configurations as a whole are not mutually complementary, and a relationship such as (2.23) or (2.26) cannot be obtained in this case either.

In conclusion, only when both of the load impedances in Fig. 5.7 are equal to $Z_0/2$ can the constant-impedance property for the input impedances be proved to be independent of the source frequency and shape of the structure.

5.4 Axially symmetric two-port self-complementary planar structures

For the purpose of investigating two-port self-complementary planar structures, a pair of geometrically self-complementary structures as shown in Figs. 5.8(a) and (b) will be considered. One port of each structure is excited by an electric source, V_{11} or V_{21}, but the other port is terminated by an impedance, Z_{l1} or Z_{l2}.

For these structures, there is no assurance of the complementary relationship, in general, if configurations as a whole are taken into account. However, these structures can be treated similarly to those in the preceding section, by replacing the limits of integration in expressions (5.22) as follows:

$$\left.\begin{aligned} Z_{l1} &= V_{12}/I_{12} = \int_f^e \boldsymbol{E}_1 \cdot \mathrm{d}\boldsymbol{l} \Big/ 2\int_h^g \boldsymbol{H}_1 \cdot \mathrm{d}\boldsymbol{l} \\ Z_{l2} &= V_{22}/I_{22} = \int_g^h \boldsymbol{E}_2 \cdot \mathrm{d}\boldsymbol{l} \Big/ 2\int_f^e \boldsymbol{H}_2 \cdot \mathrm{d}\boldsymbol{l} \end{aligned}\right\} \tag{5.27}$$

Then, the series of relationships from (5.23) to (5.26) also hold for the structures in Fig. 5.8.

Consequently, only when the load impedances Z_{l1} and Z_{l2} are equal to $Z_0/2$ can the constant-impedance property (5.26) for the input impedances be obtained as independent of the source frequency and shape of the structures, as in the case of the preceding section.

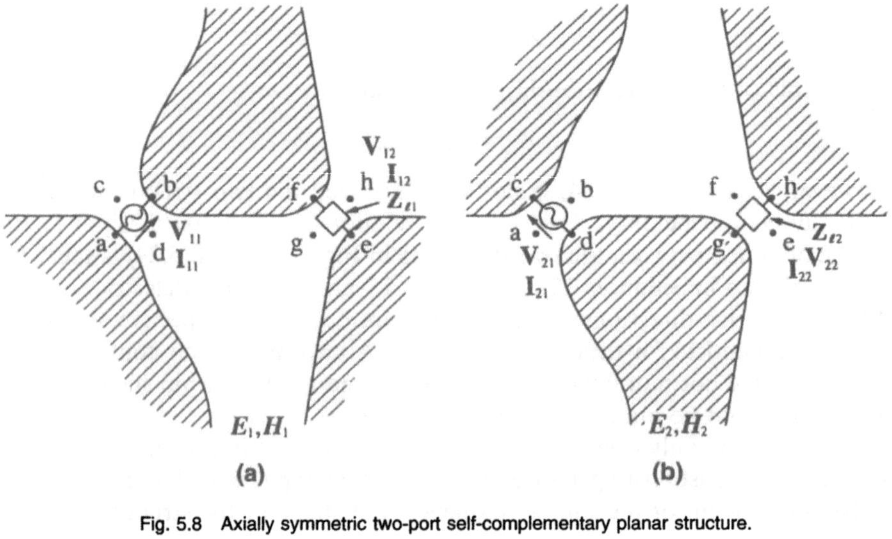

Fig. 5.8 Axially symmetric two-port self-complementary planar structure.

In connection with this conclusion, some supplementary discussion needs to be added for a better understanding of the properties of two-port complementary structures.

A simple example of the structures to be examined is shown in Fig. 5.9. The directions of the electric sources are shown by the solid arrows in the figure, and the broken arrows show the directions of the impressed magnetic fields on the front side of the conducting sheet. In general, the direction of the electric source in the complementary structure is in the direction of the magnetic field around the source in the original structure. Accordingly, the relative directions between the two electric sources, V_{11} and V_{12}, in Figs 5.9(a) and (c) are not maintained in their complementary structures (b) and (d), which is easily understood by examining the relative directions of V_{21} and V_{22} in the figures. This is also true when a phase difference is introduced between two electric sources.

This means that the two-port structures in Fig. 5.9 cannot be self-complementary, in spite of the fact that their geometrical shapes are evidently self-complementary. Consequently, no constant-impedance property can be derived for these two-port structures.

In addition, if one of their two ports is opened as shown in Fig. 5.10(a), then the corresponding port in its complementary structure becomes short-circuited as shown in Fig. 5.10(b). On the other hand, a short-circuited port in Fig. 5.10(c) corresponds to the opened port in its complementary structure as shown in Fig. 5.10(d). Accordingly, the structures mentioned above are evidently not

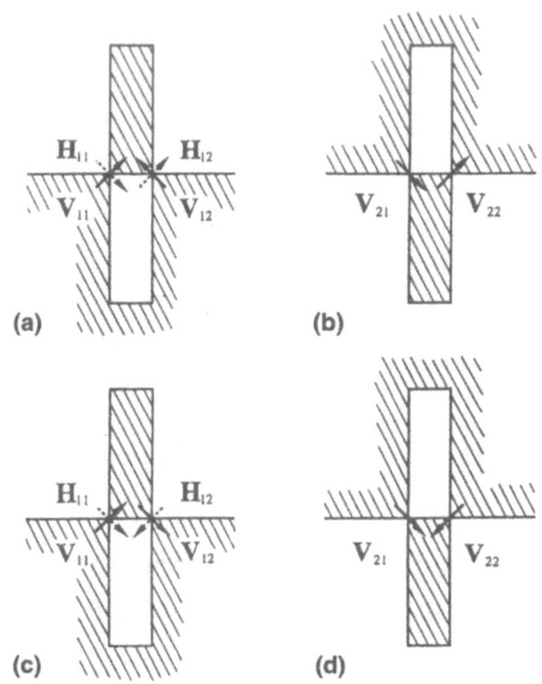

Fig. 5.9 Explanation of pseudo-self-complementary and axially symmetric two-port structures: (a) and (b) are complementary but not identical, the same is true also for (c) and (d).

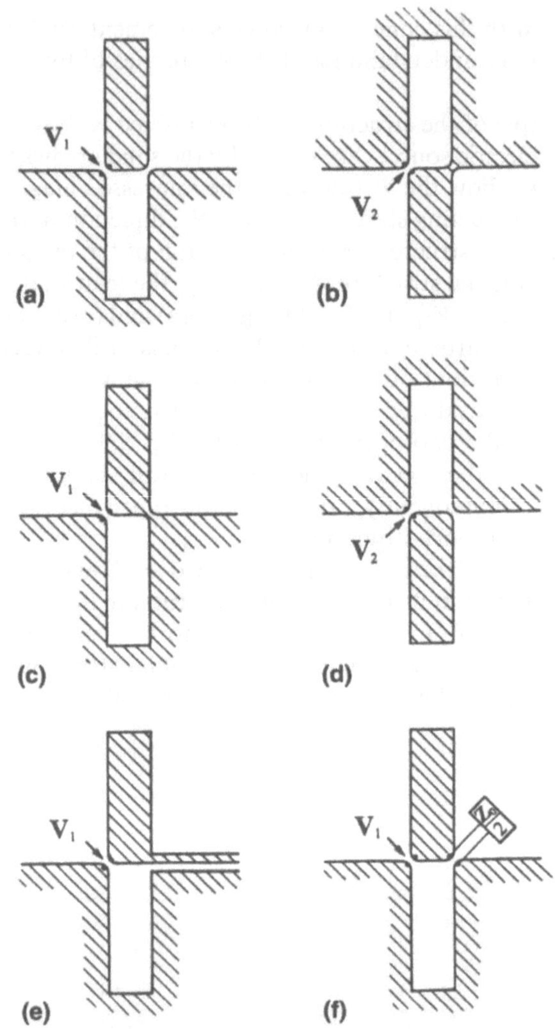

Fig. 5.10 Explanation of pseudo-self-complementary structures: (a) and (b) are complementary but not identical, the same is true also for (c) and (d), however (e) and (f) are self-complementary structures.

self-complementary, though the structures in Figs 5.10(e) and (f) are definitely self-complementary owing to their proper termination [see, for example, 5.1].

5.5 Coupling-less property between loaded unipole-notch type self-complementary structures

The shape of the unipole-notch type antenna is geometrically self-complementary, but the configuration as a whole is not necessarily self-complementary. It actually becomes self-complementary only when one of its two ports is loaded with an impedance $Z_0/2$, as explained in the preceding section. In this section, the existence of the mutual coupling between such loaded unipole-notch type self-complementary antennas will be investigated.

For that purpose, two loaded unipole-notch type self-complementary antennas as shown in Fig. 5.11(a) will be considered, where two antennas are excited with the electric source currents $\boldsymbol{J}_0$ and $\alpha \boldsymbol{J}_0$, respectively. The factor α in the source current is an arbitrary constant. The complementary structure is given in Fig. 5.11(b), where the sources are replaced by the magnetic source currents $\boldsymbol{J}_{0m}$ and $\alpha \boldsymbol{J}_{0m}$. These sources can be transformed again to the equivalent electric source currents $\boldsymbol{J}_0'$ and $\alpha \boldsymbol{J}_0'$, respectively, as shown in Fig. 5.11(c).

(a)

(b)

(c)

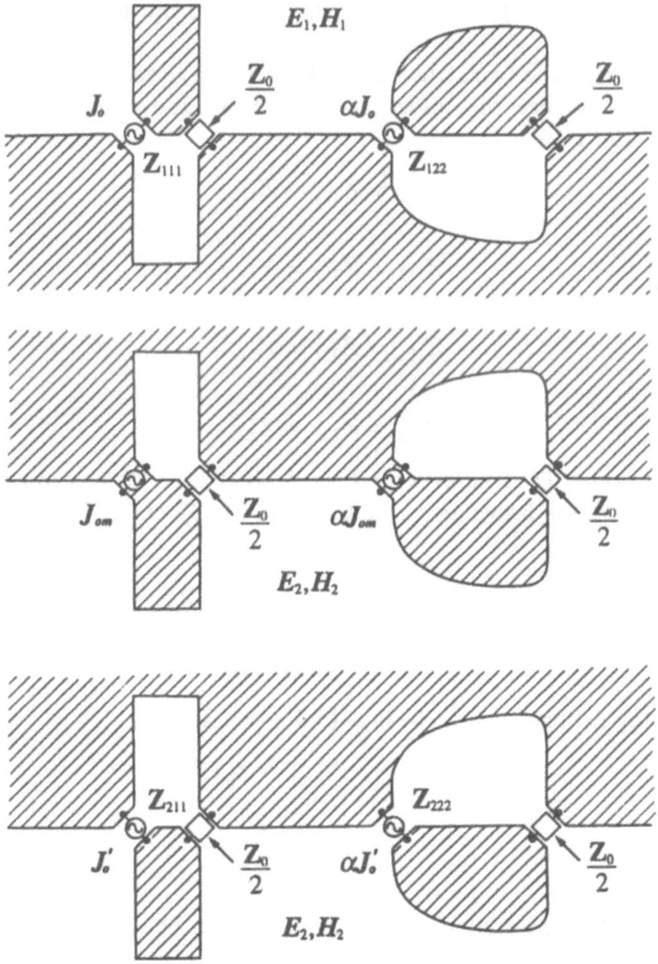

Fig. 5.11 Loaded unipole-notch type two-element self-complementary antennas: (a) and (b) are complementary, (b) and (c) are equivalent.

Now, let the electromagnetic fields for Figs 5.11(a) and (b) or (c) be $\boldsymbol{E}_1$, $\boldsymbol{H}_1$ and $\boldsymbol{E}_2$, $\boldsymbol{H}_2$, respectively, and the terminal voltages and the input currents for (a) and (c) be V_1, V_2, V_1', V_2' and I_1, I_2, I_1', I_2', respectively. Then, the terminal voltages can be expressed as:

$$\left.\begin{aligned} V_1 &= Z_{111}I_1 + Z_{112}I_2 \\ V_2 &= Z_{121}I_1 + Z_{122}I_2 \end{aligned}\right\} \tag{5.28}$$

$$\left.\begin{aligned} V_1' &= Z_{211}I_1' + Z_{212}I_2' \\ V_2' &= Z_{221}I_1' + Z_{222}I_2' \end{aligned}\right\} \tag{5.29}$$

where the Z_{irs} terms are the self-impedances and mutual impedances of these antenna systems.

By comparing the two structures in Figs 5.11(a) and (c), it is found that they are identical. Hence the impedances in (5.28) and those in (5.29) must be common, and thus

$$\left.\begin{aligned} Z_{111} &= Z_{211} = Z_{11} \\ Z_{112} &= Z_{212} = Z_{12} \\ Z_{121} &= Z_{221} = Z_{21} \\ Z_{122} &= Z_{222} = Z_{22} \end{aligned}\right\} \tag{5.30}$$

In addition, the structures in Figs 5.11(a) and (c) are mutually complementary, and an impedance relationship (3.13) exists between them. This leads to

$$[Z_{2rs}] = [Z_{1rs}]^{-1}(Z_0/2)^2 \tag{5.31}$$

$$(r = 1, 2; \quad s = 1, 2)$$

Introducing expressions (5.30) into (5.31), we find

$$[Z_{rs}] = [Z_{rs}]^{-1}(Z_0/2)^2 \tag{5.32}$$

or

$$[Z_{rs}][Z_{rs}] = [1](Z_0/2)^2 \tag{5.33}$$

where [1] is the unit matrix. This expression can be explicitly written as

$$\begin{bmatrix} Z_{11} & Z_{12} \\ Z_{21} & Z_{22} \end{bmatrix}\begin{bmatrix} Z_{11} & Z_{12} \\ Z_{21} & Z_{22} \end{bmatrix} = \begin{bmatrix} 1 & 0 \\ 0 & 1 \end{bmatrix}(Z_0/2)^2 \tag{5.34}$$

Decomposing this matrix expression into four relationships, we obtain

$$\left.\begin{aligned} Z_{11}^2 + Z_{12}Z_{21} &= (Z_0/2)^2 \\ Z_{21}Z_{12} + Z_{22}^2 &= (Z_0/2)^2 \\ (Z_{11} + Z_{22})Z_{12} &= 0 \\ (Z_{11} + Z_{22})Z_{21} &= 0 \end{aligned}\right\} \tag{5.35}$$

From the condition

$$Z_{11} + Z_{22} \neq 0 \tag{5.36}$$

we finally obtain

$$\left.\begin{aligned} Z_{12} = Z_{21} &= 0 \\ Z_{11} = Z_{22} &= Z_0/2 \end{aligned}\right\} \tag{5.37}$$

Consequently, it is found that the loaded unipole-notch type two-element self-complementary antennas shown in Fig. 5.11 have no mutual coupling between the two elements that is independent of the source frequency, their shapes, and the spacing between the two elements. At the same time, the input impedances of the two respective elements are always constant, irrespective of the existence of the other element and the magnitudes of their exciting currents.

Next, the general case of an n-element system of a loaded unipole-notch type self-complementary antenna will be considered. In this case the matrix representations of the terminal voltages corresponding to relationships (5.28) and (5.29) are given as

$$[V_r] = [Z_{1rs}][I_s] = [Z_{1rs}][\alpha_s]I_1 \tag{5.38}$$

$$(r = 1, 2, 3, \cdots, n; \quad s = 1, 2, 3, \cdots, n)$$

$$[V_r'] = [Z_{2rs}][I_s'] = [Z_{2rs}][\alpha_s]I_1' \tag{5.39}$$

for two complementary antenna systems, where

$$I_s = \alpha_s I_1 \tag{5.40}$$

and α_2, α_3, ... are independent arbitrary constants with $\alpha_1 = 1$. Furthermore, these two antenna systems are identical, and the impedances in (5.38) and (5.39) must have common values, so we can write

$$[Z_{1rs}] = [Z_{2rs}] = [Z_{rs}] \tag{5.41}$$

Hence we have

$$V_r/I_r = V_r'/I_r' = (Z_{r1} + \alpha_2 Z_{r2} + \alpha_3 Z_{r3} + \cdots + \alpha_r Z_{rr} + \cdots + \alpha_n Z_{rn})/\alpha_r \tag{5.42}$$

These ratios of voltages and currents for the corresponding two terminals of mutually complementary structures can be expressed respectively by the integrations of the electromagnetic fields using the same expressions as given in (5.21). Accordingly, we find

$$(V_r/I_r)(V_r'/I_r') = (Z_0/2)^2 \tag{5.43}$$

From (5.42) and (5.43) we obtain

$$(Z_{r1} + \alpha_2 Z_{r2} + \alpha_3 Z_{r3} + \cdots + \alpha_r Z_{rr} + \cdots + \alpha_n Z_{rn})/\alpha_r = Z_0/2 \tag{5.44}$$

This expression holds for any arbitrary values of α_r , hence the relationships

$$\left.\begin{aligned} Z_{rr} &= Z_0/2 \\ Z_{rs} &= 0 \quad (r \neq s) \end{aligned}\right\} \tag{5.45}$$

correspond to the general case of (5.37).

Results (5.45) can be checked from another stand-point. Matrix expression (5.33) is the relationship obtained for the case of two elements, but an identical expression must be derived using the same procedure, also for the case of n elements. Thus

$$[Z_{rs}][Z_{rs}] = [1](Z_0/2)^2 \tag{5.46}$$

and if we introduce the relationship

$$Z_{rs} = 0 \quad (r \neq s) \tag{5.47}$$

into (5.46), then this reduces to

$$Z_{rr}^2 = (Z_0/2)^2$$

and we obtain

$$Z_{rr} = Z_0/2 \qquad (r = 1, 2, 3, \cdots, n) \tag{5.48}$$

Consequently, relationships (5.45) are proved to be sufficient as the solution for a general case of n elements.

As an example of the application of the coupling-less property between the loaded unipole-notch type elements described above, the four-element antenna system shown in Fig. 5.12(a) will be considered. The input impedances of these four elements are always $Z_0/2$, irrespective of the existence of other elements. This is a natural consequence of the coupling-less property of these elements. Accordingly, the load impedance for the fed element can be replaced by the element located second from the left, by connecting through a transmission line with the characteristic impedance $Z_0/2$. Likewise, the load impedance for the second element can be replaced by the third element connected by a feed line. Thus, all the four elements can be connected in tandem as shown in Fig. 5.12(b). This structure is a general case of the axially symmetric self-complementary antenna, which has already been shown in Fig. 4.4.

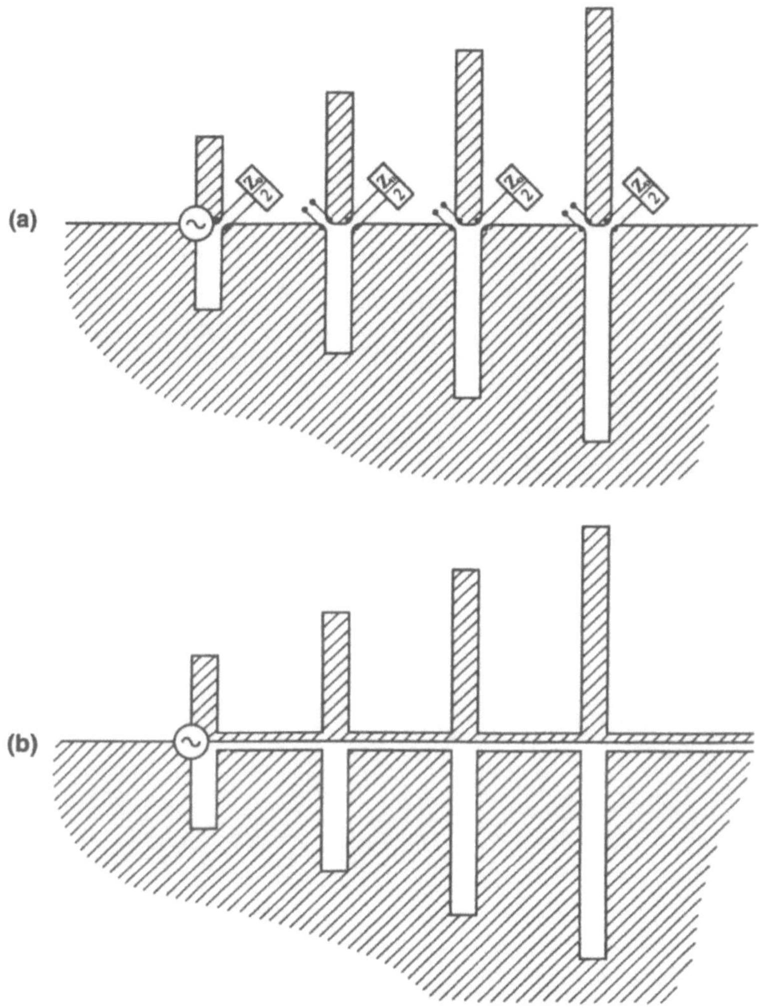

Fig. 5.12 Tandem connection of loaded unipole-notch type antennas.

Incidentally, the coupling-less property between loaded unipole-notch type self-complementary planar structures has recently been experimentally confirmed by T. Kasahara.

6 • THREE-DIMENSIONAL SELF-COMPLEMENTARY STRUCTURES

6.1 A pair of dual structures consisting of crossed infinite planar sheets of compound perfect conductors

A cross of infinite planar sheets of compound perfect conductors, as shown in Fig. 6.1(a) will now be considered, where the crossing angle is rectangular and the shapes of the perfect electric conductor (PEC) and the perfect magnetic conductor (PMC) are both symmetrical with respect to the crossing axis. Furthermore, in the two mutually perpendicular infinite sheets, the shapes of the portions of the PEC and the PMC are identical, but their properties are interchanged. In such a structure, the whole space is divided into four perfectly shielded and independent partial spaces. For these four partial spaces, we assume four symmetrical electric source currents $\boldsymbol{J}_{01}$, $\boldsymbol{J}_{02}$, $\boldsymbol{J}_{03}$ and $\boldsymbol{J}_{04}$, respectively.

Next, let the electric and magnetic properties in the structure of Fig. 6.1(a) be interchanged, producing another structure as shown in Fig. 6.1(b), where the symmetrical electric source currents are replaced by the anti-symmetrical magnetic source currents, $\boldsymbol{J}_{0m1}$, $\boldsymbol{J}_{0m2}$, $\boldsymbol{J}_{0m3}$ and $\boldsymbol{J}_{0m4}$. The magnitudes of these currents are assumed to be in the following relationships:

$$\left.\begin{aligned} \boldsymbol{J}_{0m1} &= \boldsymbol{J}_{01}, \qquad & \boldsymbol{J}_{0m2} &= -\boldsymbol{J}_{02} \\ \boldsymbol{J}_{0m3} &= \boldsymbol{J}_{03}, \qquad & \boldsymbol{J}_{0m4} &= -\boldsymbol{J}_{04} \end{aligned}\right\} \tag{6.1}$$

Thus, Figs 6.1(a) and (b) form a pair of dual structures. Accordingly, the electromagnetic fields, $\boldsymbol{E}_1$, $\boldsymbol{H}_1$ for (a) and $\boldsymbol{E}_2$, $\boldsymbol{H}_2$ for (b), are expressed in the duality relationships given by (2.24), that is

$$\boldsymbol{E}_2 = \mp \boldsymbol{H}_1, \qquad \boldsymbol{H}_2 = \pm \gamma \boldsymbol{E}_1 \tag{6.2}$$

The double signs here show that the sign must be changed for the two sides of every semi-infinite wing of compound perfect conductors. The upper signs apply for the partial spaces where the two kinds of source current are in the same direction for the mutually dual structures.

(a)

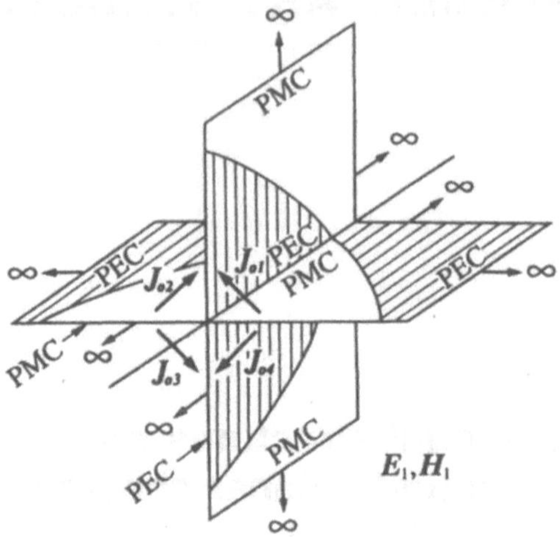

(b)

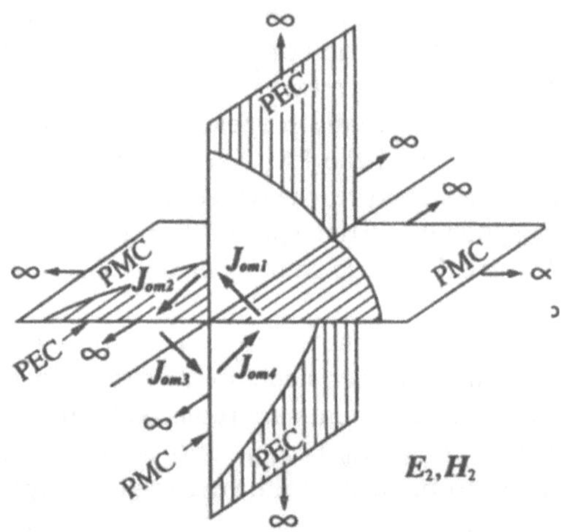

Fig. 6.1 Crossed infinite planar sheets of compound perfect conductors: (a) and (b) are dual structures.

6.2 Three-dimensional complementary structures

Three-dimensional mutually complementary structures can be easily derived from the dual structures described in the preceding section, by eliminating the fictitious PMC sheets from them. For that purpose, the symmetries of the electromagnetic fields for the structures in Figs 6.1(a) and (b) will be examined, and it will be found that the boundary conditions for the PMC surface are automatically satisfied, as in the case of two-dimensional planar structures as discussed in section 2.3. Therefore, all the PMC sheets in Figs 6.1(a) and (b) can be taken away without disturbing any electromagnetic fields. By eliminating such PMC sheets, we obtain the PEC structures as shown in Figs 6.2(a) and

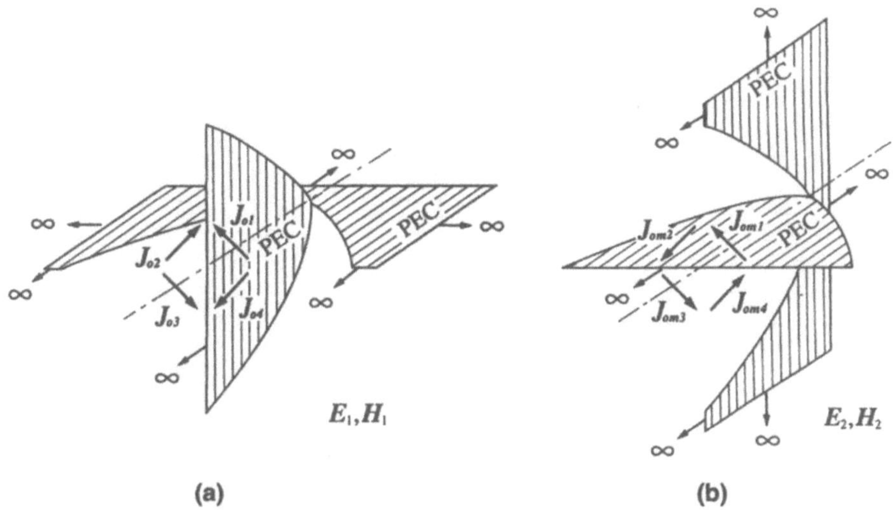

Fig. 6.2 Crossed semi-infinite planar sheets of perfect electric conductor: semi-infinite conducting sheets are complementary in (a) and (b), and the structures in (a) and (b) are mutually complementary.

(b). The electromagnetic fields for these two structures must also be as given in relationship (6.2).

As is apparent from the derivation procedure, the PEC structures in Fig. 6.2 are mutually complementary. In addition, the mutually perpendicular PEC sheets in the respective structure (a) or (b) are also mutually complementary.

6.3 Three-dimensional self-complementary structures

The excitations for the structures shown in Fig. 6.2 are generalized source distributions. As a particular case of the excitation, consider the example illustrated in Fig. 6.3(a), where an electric source is connected between two terminals located on two mutually complementary portions of an infinite PEC planar sheet. The shapes of both portions are symmetrical with respect to the crossing axis, and these two portions are twisted with respect to each other by 90° around the crossing axis, as illustrated in Fig. 6.3.

The electric source here can be divided into four equal partial currents, and they are assumed to be included in the four respective quarter-spaces around the crossing axis. Hence, these partial currents constitute the symmetrical electric source currents, and such a structure is evidently a particular case of Fig. 6.2(a).

In order to facilitate further analysis, a detailed illustration of the feed points in a quarter-space is shown in Fig. 6.4(a). Its complementary structure with the

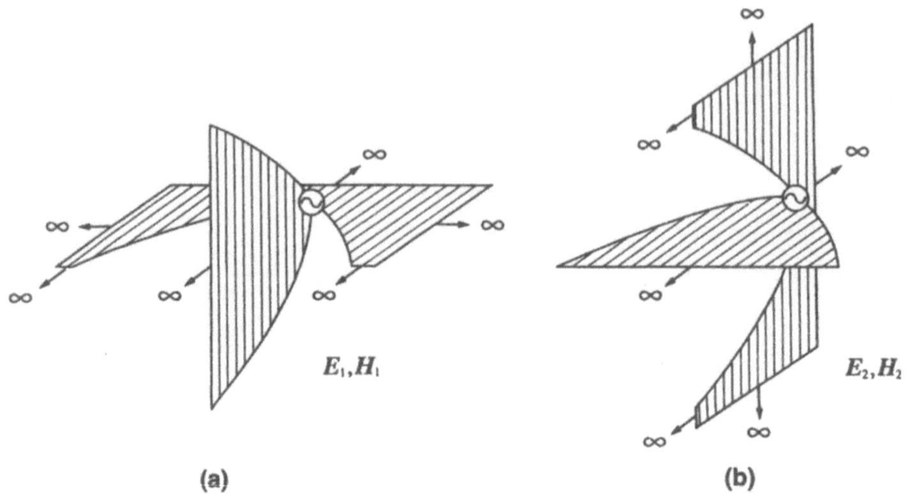

Fig. 6.3 Three-dimensional self-complementary structures: (a) and (b) are complementary and identical.

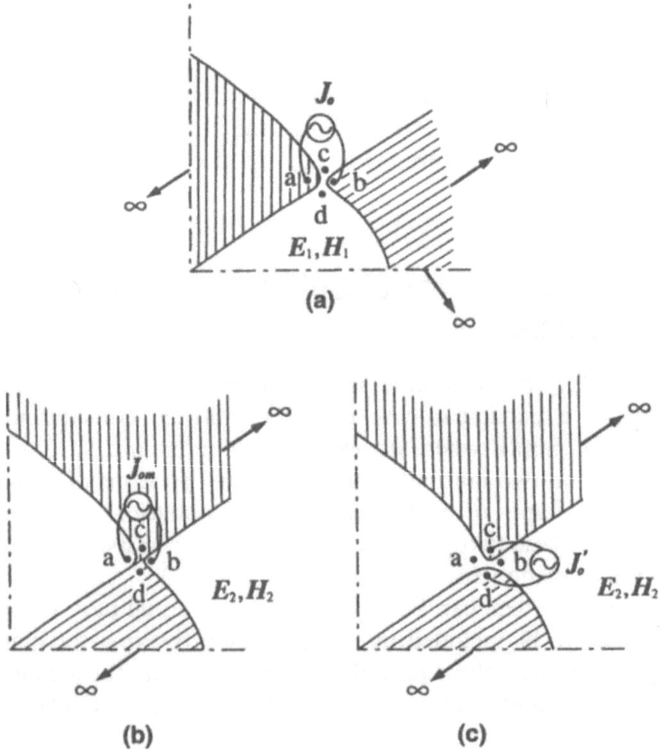

Fig. 6.4 Magnified illustrations of quarter-spaces for the structures in Fig. 6.3: (a) and (b) are complementary, (b) and (c) are equivalent.

magnetic source current is shown in Fig. 6.4(b), which is equivalent to the structure in Fig. 6.4(c). By comparing the structures in Figs 6.4(b) and 6.2(b), it is easily understood that the former structure is a quarter-space of the latter. Also, by comparing the structures in Figs 6.4(c) and 6.3(b), we find that the former structure corresponds to a quarter-space of the latter.

Therefore, the structures in Figs 6.3(a) and (b) are mutually complementary, and their electromagnetic fields, $\boldsymbol{E}_1$, $\boldsymbol{H}_1$ and $\boldsymbol{E}_2$, $\boldsymbol{H}_2$, related as expressed in (6.2). Furthermore, it is easily understood from their derivation procedure that the configurations of the structures in (a) and (b) are wholly identical. This means that the structures in Figs 6.3(a) and (b) are self-complementary. Thus, we finally obtain a three-dimensional self-complementary structure.

Now, the input impedance of the three-dimensional self-complementary structure can be discussed by considering the quarter-space illustrated in Fig. 6.4. Let the partial input impedances for the quarter-spaces in Figs 6.4(a) and (c) be respectively Z_{p1} and Z_{p2}, and their values are given by the ratios of the integrations for the two electromagnetic fields [1.5, 1.6] as follows:

$$\left.\begin{aligned} Z_{p1} &= \int_b^a \boldsymbol{E}_1 \cdot \mathrm{d}\boldsymbol{l} \Bigg/ \int_c^d \boldsymbol{H}_1 \cdot \mathrm{d}\boldsymbol{l} \\ Z_{p2} &= \int_d^c \boldsymbol{E}_2 \cdot \mathrm{d}\boldsymbol{l} \Bigg/ \int_b^a \boldsymbol{H}_2 \cdot \mathrm{d}\boldsymbol{l} \end{aligned}\right\} \tag{6.3}$$

These expressions are similar to those in (3.2), which were obtained for the structures consisting of a single sheet of PEC, but in expressions (6.3) the factor "2" does not appear, because the integration of the magnetic field must be performed only in the quarter-space.

Since this antenna is self-complementary, as mentioned above, Z_{p1} and Z_{p2} have a common value, Z_p, hence from (6.2) and (6.3) we find that

$$Z_p^2 = Z_{p1} Z_{p2} = 1/\gamma = Z_0^2 \tag{6.4}$$

which leads to

$$Z_p = Z_0 \tag{6.5}$$

for a quarter-space. However, the total current for the structure in Fig. 6.3 is four times larger than the current in a quarter-space, while the terminal voltages are common between them.

Consequently, the input impedance Z of the three-dimensional four-wing self-complementary structure is given by

$$Z = Z_0/4 \simeq 30\pi \quad [\Omega] \tag{6.6}$$

In general, a $2N$-wing three-dimensional self-complementary structure can be treated similarly, and the input impedance Z_N is given as

$$Z_N = Z_0/2N \simeq 60\pi/N \quad [\Omega] \tag{6.7}$$

Relationships (6.6) and (6.7) show that the constant-impedance property is proved also for this type of three-dimensional self-complementary structure.

6.4 Examples of three-dimensional self-complementary structures

One of the simplest examples of a three-dimensional self-complementary structure is shown in Fig. 6.5(a).

If the transverse dimensions of the strip and the slot in that structure are sufficiently small compared with the wavelength, then this structure works as a non-radiative transmission line with the characteristic impedance $Z_0/4$. This is one type of self-complementary transmission line. It is interesting to see the difference between the two self-complementary transmission lines in Figs 6.5(a) and 4.5(a), where the characteristic impedances are respectively $Z_0/4$ and $Z_0/2$. Also, it is understood that the infinitely long portion of the structure in Fig. 6.5(a) can be substituted by a load impedance $Z_0/4$ as shown in Fig. 6.5(b), in a similar way to the case of the structure in Fig. 4.5(b). Such a substitution does not cause any appreciable variations in the electromagnetic fields, except in the vicinity of the truncated portion of the strip line.

Another simple example is shown in Fig. 6.6, where the widths of the strips and the slots are assumed to be sufficiently small compared with the wavelength. Hence, the infinitely long strip and the infinitely long slot can be substituted by a load impedance $Z_0/4$, as explained above. With such a substitution, the structure in Fig. 6.6 reduces to a pair of monopole-slot antennas mounted in opposite directions to each other on both sides of an infinite ground plane.

A typical example of a three-dimensional self-complementary antenna is shown in Fig. 6.7.

The bandwidth of the radiation characteristics of this antenna is much broader than that for Fig. 6.6, though the input impedances for both antennas

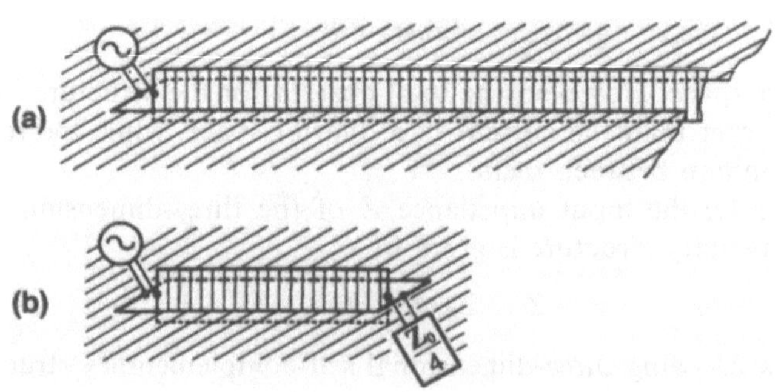

Fig. 6.5 Self-complementary transmission line consisting of a strip line and a slotted infinite planar sheet of perfect conductor: (a) without termination, (b) with termination (both $Z \simeq 30\pi\ \Omega$).

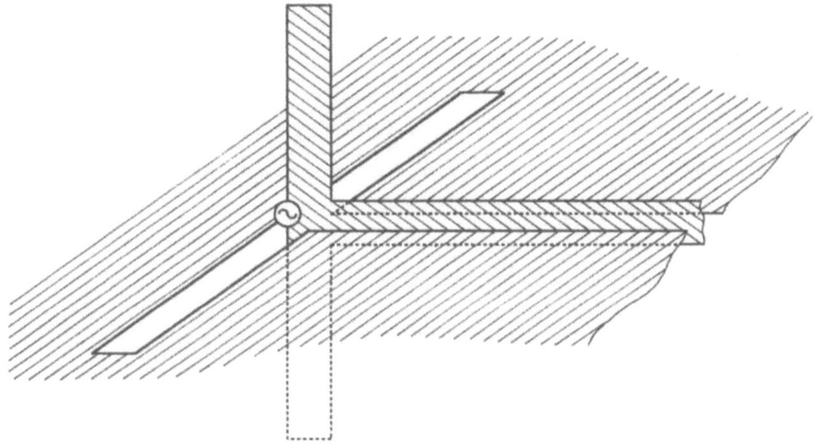

Fig. 6.6 A simple type of three-dimensional self-complementary antenna ($N = 2$, $Z \simeq 30\pi\ \Omega$).

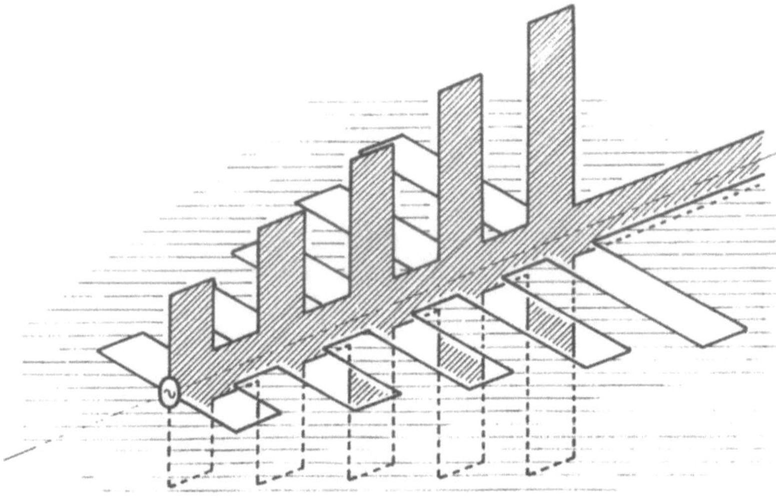

Fig. 6.7 An example of a three-dimensional self-complementary antenna ($N = 2$, $Z \simeq 30\pi\ \Omega$).

are always $Z_0/4$, independently of the source frequency and the shape of the structures. Some related discussions will be made later in Chapter 11.

7 • STACKED SELF-COMPLEMENTARY ANTENNAS

7.1 Stacking of self-complementary antennas

In order to improve the radiation characteristics of antennas, especially to increase their directivities or power gains, the stacked antenna technique is often utilized effectively in practice. In such cases, there are mutual interactions between the element antennas, mainly owing to the effects of the mutual impedances. Accordingly, the constant-impedance property of self-complementary antennas is not necessarily preserved when they are introduced as element antennas into a stacked antenna. However, the author has proposed various types of stacked self-complementary antenna, where all the element antennas have constant impedance, independently of the source frequency and their shapes.

In this chapter, various conceivable types of stacked self-complementary antenna systems are presented, and their configurations as a whole, including their feeding methods, are examined. It is found from the results of these studies that some of these types are actually self-complementary as a whole, and also have the constant-impedance property at each feed point, although others do not possess such properties. This means that not all structures that are geometrically self-complementary will invariably be self-complementary in stacked antenna systems.

Some proposed ways of stacking are as follows:

1 Co-planar stacking of self-complementary antennas [1.7, 1.8].
   (a) Periodic stacking
   (b) Alternate periodic stacking
2 Side-by-side stacking of self-complementary antennas [1.7,1.8].
   (a) Periodic stacking
   (b) Alternate periodic stacking
3 Compound-stacking of self-complementary antennas [1.8, 1.9].

In addition, some variations of these arrangements will also be discussed in this chapter.

7.2 Co-planar stacked self-complementary antennas

In order to explain the principle of co-planar stacked self-complementary antennas, let us consider a co-planar periodic structure as shown in Fig. 7.1(a), where a unipole-notch antenna with tail line is mounted on each PEC parallel belt. In this structure, the unipole-notch antenna on the PEC belt is not self-complementary, in the strict sense, when such a single element is isolated in free space, because this antenna becomes self-complementary only when it is mounted on a semi-infinite PEC planar sheet.

As a natural consequence of the property of periodic structure of Fig. 7.1(a), all the electric sources for the element structures must be identical. For this reason, detailed discussions will be continued by selecting just one of the electric sources, as shown in Fig. 7.2(a).

The electric current here can be divided into two equal portions, where one portion is on the front side of the conducting sheet belt and the other portion is

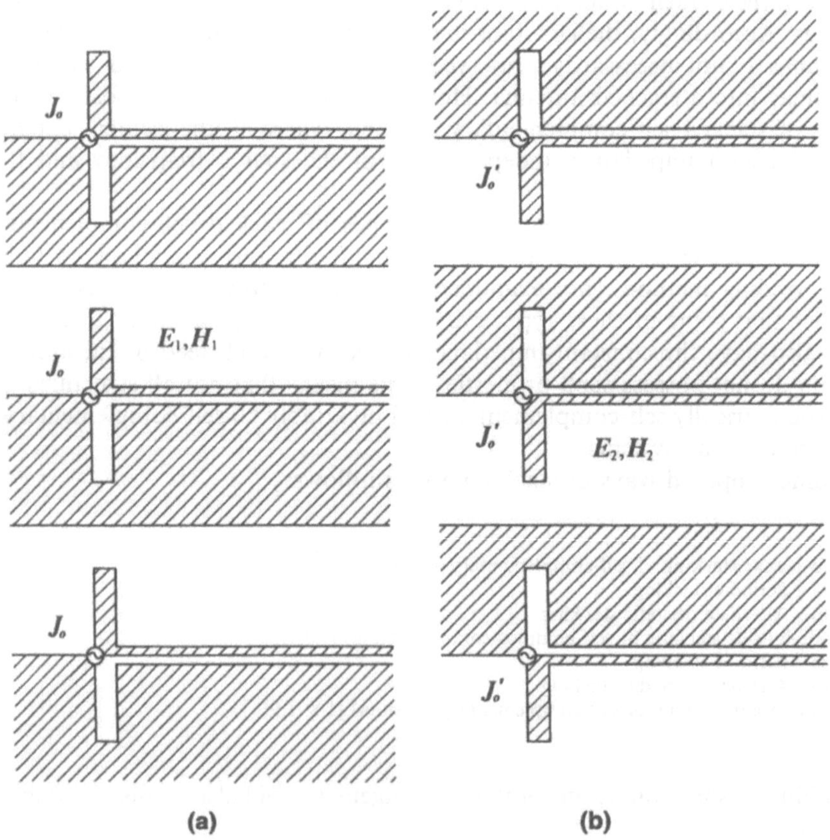

Fig. 7.1 Periodically stacked co-planar antennas: (a) and (b) are complementary and identical.

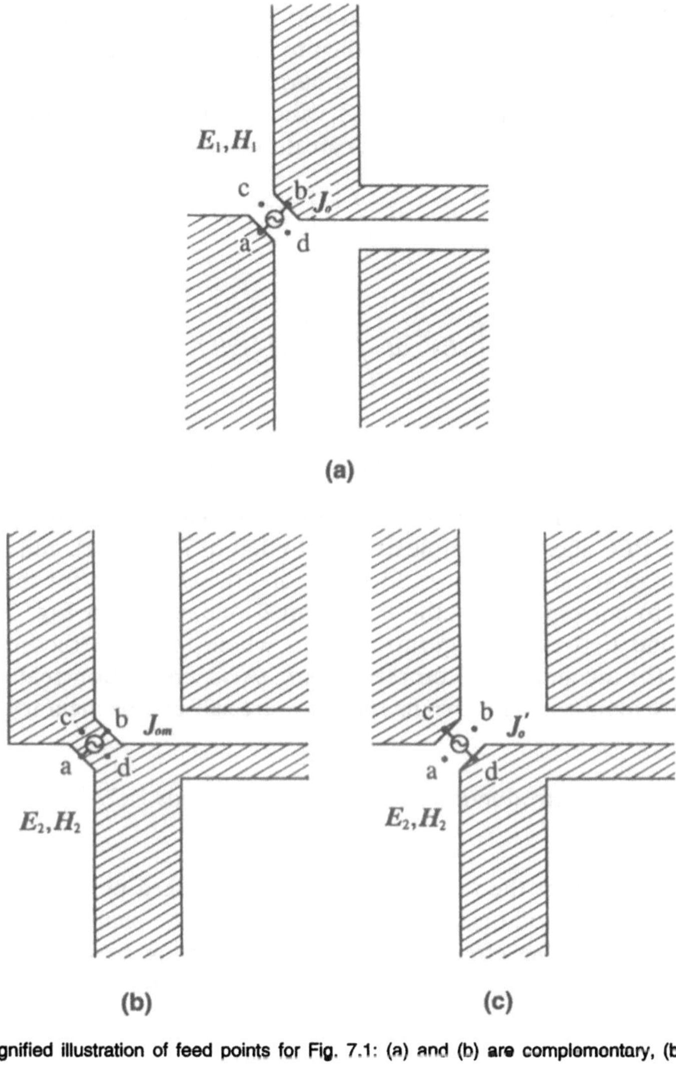

Fig. 7.2 Magnified illustration of feed points for Fig. 7.1: (a) and (b) are complementary, (b) and (c) are equivalent.

on the reverse side. These electric currents together make symmetrical electric source currents. Accordingly, the structure that is complementary to Fig. 7.2(a) is given in Fig. 7.2(b), where the excitation sources are transformed to the anti-symmetrical magnetic source currents. Consequently, the electromagnetic fields $\boldsymbol{E}_1$, $\boldsymbol{H}_1$ for the structure in Fig. 7.2(a), and $\boldsymbol{E}_2$, $\boldsymbol{H}_2$ for the structure in Fig. 7.2(b), are in the duality relationships given by (2.24), that is

$$\boldsymbol{E}_2 = \mp \boldsymbol{H}_1 \qquad \text{and} \qquad \boldsymbol{H}_2 = \pm \gamma \boldsymbol{E}_1 \tag{7.1}$$

where the double signs are explained in Chapter 2, section 2.3.

Now, the magnetic source currents in Fig. 7.2(b) circulate from terminal "a" to terminal "b" on the front side and then to "a" on the back side. This means that they are equivalent to the electric current source shown in Fig. 7.2(c). Accordingly, the input impedances Z_1 and Z_2 for the structures in Figs 7.2(a) and (c) are given by the same expressions as (3.2), that is

$$\left.\begin{aligned} Z_1 &= \int_b^a \boldsymbol{E}_1 \cdot d\boldsymbol{l} \bigg/ 2\int_c^d \boldsymbol{H}_1 \cdot d\boldsymbol{l} \\ Z_2 &= \int_d^c \boldsymbol{E}_2 \cdot d\boldsymbol{l} \bigg/ 2\int_b^a \boldsymbol{H}_2 \cdot d\boldsymbol{l} \end{aligned}\right\} \tag{7.2}$$

and hence we obtain

$$Z_1 Z_2 = 1/4\gamma = (Z_0/2)^2 \tag{7.3}$$

Next, let us consider a periodic structure as shown in Fig. 7.1(b), where the shape of the PEC sheets is complementary to that of Fig. 7.1(a), and all of their electric sources are also identical. Furthermore, the excitation for the structure in Fig. 7.1(b) corresponds to the transformed one shown in Fig. 7.2(c), and hence we find that the configurations as a whole for the structures in Figs 7.1(a) and (b) are mutually complementary. Also, inspection readily reveals that they are identical. This means that they are self-complementary and so

$$Z_1 = Z_2 \tag{7.4}$$

Consequently, the introduction of (7.4) into (7.3) leads [1.7, 1.8] to

$$Z_1 = Z_2 = Z_0/2 \tag{7.5}$$

This result shows that the input impedance at each feed point is always constant, independently of the source frequency and the shape of the element antennas, including the effects of the mutual coupling between the element antennas. Thus, the constant-impedance property is also proved for periodically stacked co-planar self-complementary antennas.

7.3 Some variations of co-planar stacked self-complementary structures

As an example of the variations of co-planar periodically stacked self-complementary structures, an alternate periodic structure is proposed as shown in Fig. 7.3(a). The alternate periodic electric source currents are assumed to be as indicated in the figure by $\boldsymbol{J}_0$ and $\alpha \boldsymbol{J}_0$, where α is an arbitrary constant, and it may include phase difference.

For the purpose of confirming the self-complementary nature of this structure, another structure which is complementary to it is considered in Fig. 7.3(b), where the excitation sources are the equivalent electric currents $\boldsymbol{J}_0'$

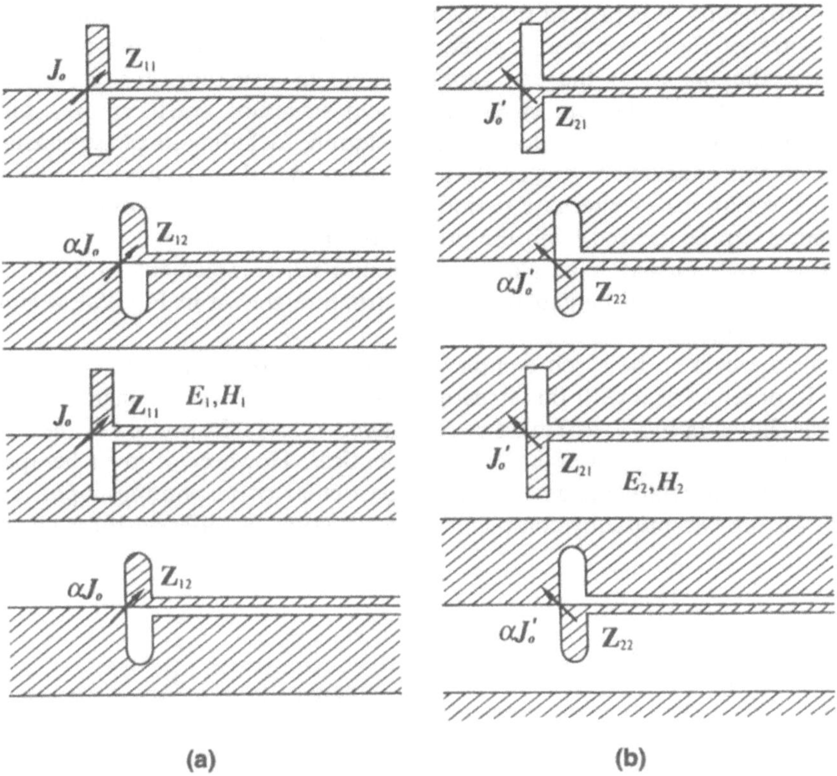

Fig. 7.3 Alternately periodic co-planar stacked antenna: (a) and (b) are complementary and identical.

and $\alpha \boldsymbol{J}_0'$, which are transformed from the anti-symmetrical magnetic source currents. Hence, the input impedance at each terminal can be expressed by the ratio of the integrations of the electromagnetic fields in a similar manner to those in (7.2). Therefore, the input impedances Z_{11}, Z_{12} for Fig. 7.3(a) and Z_{21}, Z_{22} for Fig. 7.3(b) are in the same relationship as (7.3), namely

$$\left.\begin{aligned} Z_{11}Z_{21} &= (Z_0/2)^2 \\ Z_{12}Z_{22} &= (Z_0/2)^2 \end{aligned}\right\} \tag{7.6}$$

Now, a careful comparison between the two structures mentioned above, especially on their relative geometrical arrangements, shows that these two structures are identical, including their excitation arrangements. Hence they are self-complementary and the corresponding impedances are respectively identical:

$$Z_{11} = Z_{21} \qquad \text{and} \qquad Z_{12} = Z_{22} \tag{7.7}$$

Introduction of (7.7) into (7.6) leads to

$$Z_{11} = Z_{21} = Z_{12} = Z_{22} = Z_0/2 \tag{7.8}$$

Consequently, it is proved that the input impedances for all terminals of the structures in Fig. 7.3 are always constant, independently of the source frequency, the shapes of the two types of element structure, and the relative geometry between the two types of element structure. Furthermore, this is also true for any arbitrary value of the constant α, including its phase constant.

From this conclusion, it is easily understood that the various forms of any further variations of alternate periodic co-planar stacked self-complementary antennas can be derived from the procedure discussed above.

Next, an unusual example of a pseudo-self-complementary co-planar stacked structure will be explained. The structure shown in Fig. 7.4(a) is just such an example.

This appears to be a self-complementary structure, although actually it is not. In order to deal with this structure, a geometrically complementary structure as shown in Fig. 7.4(b) must be considered. Careful inspection of these two structures seems to show that they are identical. However, it must be appreciated that the relative geometry of any particular corresponding input terminals for the two mutually complementary structures is different for each case.

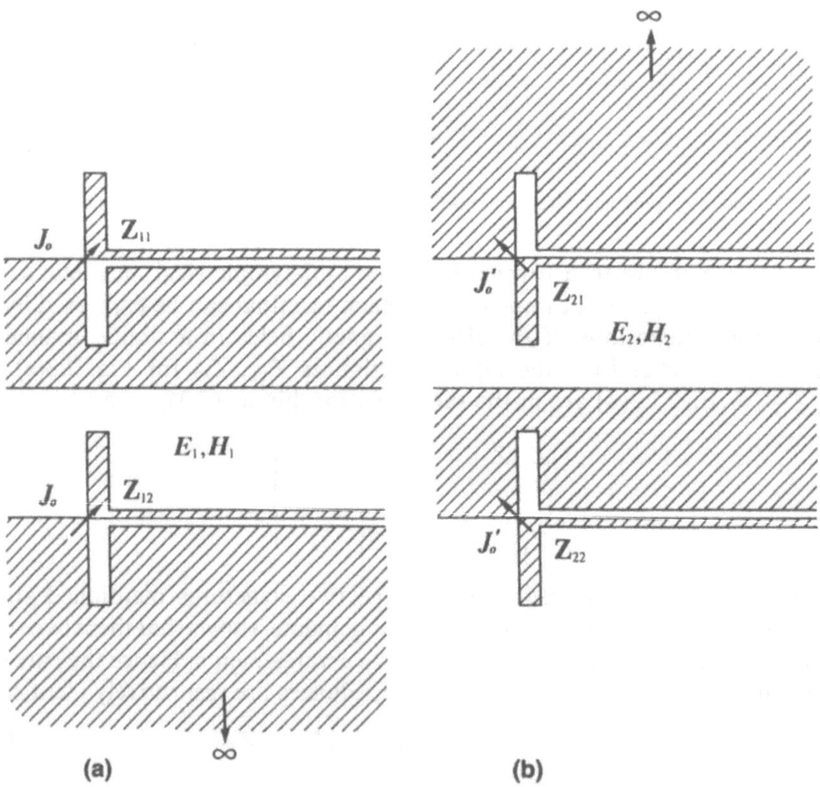

Fig. 7.4 Pseudo-self-complementary co-planar stacked structures: (a) and (b) are complementary but relative constitutions are not identical.

Therefore, the relationships between the input impedances of two mutually complementary structures are given by

$$Z_{12} = Z_{21} \qquad \text{and} \qquad Z_{11} = Z_{22} \tag{7.9}$$

where no impedance relationship between mutually complementary input terminals is obtained.

Also, from the complementary nature of these two structures we obtain

$$\left.\begin{aligned} Z_{11}Z_{21} &= (Z_0/2)^2 \\ Z_{12}Z_{22} &= (Z_0/2)^2 \end{aligned}\right\} \tag{7.10}$$

Unfortunately, combinations of (7.9) and (7.10) do not provide any additional relationship. This is a natural consequence of the non-self-complementary nature of the structure in Fig. 7.4(a) or (b), although the structure as a whole appears to be self-complementary. Such a structure is termed a "pseudo-self-complementary structure" in this book.

7.4 Side-by-side stacked self-complementary antennas

In general, side-by-side stacking of element antennas is very effective for improving the radiation characteristics of antenna systems in practice. In this section, several methods of stacking for self-complementary structures will be investigated.

For this purpose, a unipole-notch antenna on a semi-infinite PEC planar sheet, as shown in Fig. 7.5(a), is considered as a simple example of an element antenna. An exciting electric current $2\boldsymbol{J}_0$ is assumed there. Next, this current can be divided into two equal portions, located on opposite sides of the PEC planar sheet. In the following discussions, the structure and the excitation illustrated in Fig. 7.5(a) will be symbolized by the greatly simplified figure shown in Fig. 7.5(b). Similarly, Fig. 7.5(c) will be symbolized by Fig. 7.5(d). In this connection, it should be noted that the declination angles of the arrows of the symbols in (b) and (d) are different, corresponding to the difference in the directions of the exciting currents $2\boldsymbol{J}_0$ in (a) and $2\boldsymbol{J}_0'$ in (c).

By using such symbols, a side-by-side periodically stacked antenna system of unipole-notch antennas can be shown as in Fig. 7.6(a). By inspecting this figure, it is easily understood that the entire configuration of this structure, including the electric source current $\boldsymbol{J}_0$ terms, are always symmetrical with respect to any plane of the element structures. Accordingly, semi-infinite PMC sheets that are complementary to the original PEC structures of the element antennas can be introduced, without any appreciable difference being made to the electromagnetic fields, and this process leads to the structure shown in Fig. 7.6(b). Thus a structure is obtained which divides the whole space into independent slab-shaped partial spaces stacked in parallel.

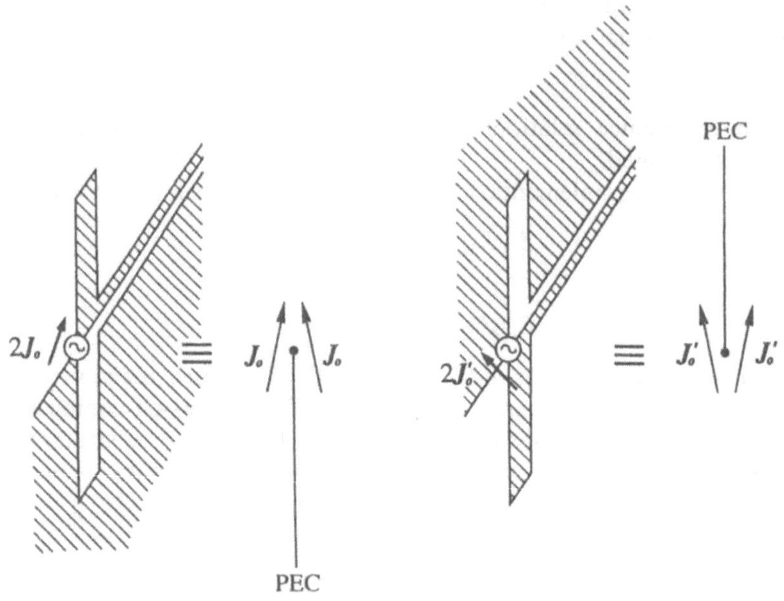

Fig. 7.5 Illustration of the symbolized sketches of the self-complementary element structures: the structures (a) and (c) are shown by symbolized sketches at (b) and (d), respectively.

Next, consider another structure as shown in Fig. 7.6(c), which is the dual of the structure in Fig. 7.6(b). In these two structures, the electric and magnetic properties are interchanged. Furthermore, the signs of the electromagnetic fields are reversed alternately at every adjacent parallel partial space so as to form anti-symmetrical magnetic source currents, $\boldsymbol{J}_{0m}$. Such a technique can be applied in this treatment because the partial spaces are perfectly shielded by the compound perfect conductors.

On the other hand, the original electromagnetic fields $\boldsymbol{E}_1, \boldsymbol{H}_1$ are transformed to $\boldsymbol{E}_2, \boldsymbol{H}_2$, and from the duality relationship between these two fields we obtain

$$\boldsymbol{E}_2 = \mp \boldsymbol{H}_1 \qquad \text{and} \qquad \boldsymbol{H}_2 = \pm \gamma \boldsymbol{E}_1 \tag{7.11}$$

where the upper and the lower signs apply to the partial spaces indicated alternately by the plus and minus signs in Fig. 7.6(c).

Now, the anti-symmetrical magnetic source currents can be replaced with the equivalent electric source current terms $\boldsymbol{J}_0'$, and the structure in Fig. 7.6(c) transforms to that in (d). Hence, the entire distributions of these electric source currents are always symmetrical with respect to any plane of the element structures. Accordingly, all the fictitious structures of the PMC sheets can be removed from Fig. 7.6(d) without disturbing any electromagnetic fields. This finally leads to the PEC structures shown in Fig. 7.6(e), where the electric source current terms, $\boldsymbol{J}_0'$, are alternately reversed as illustrated in the figure.

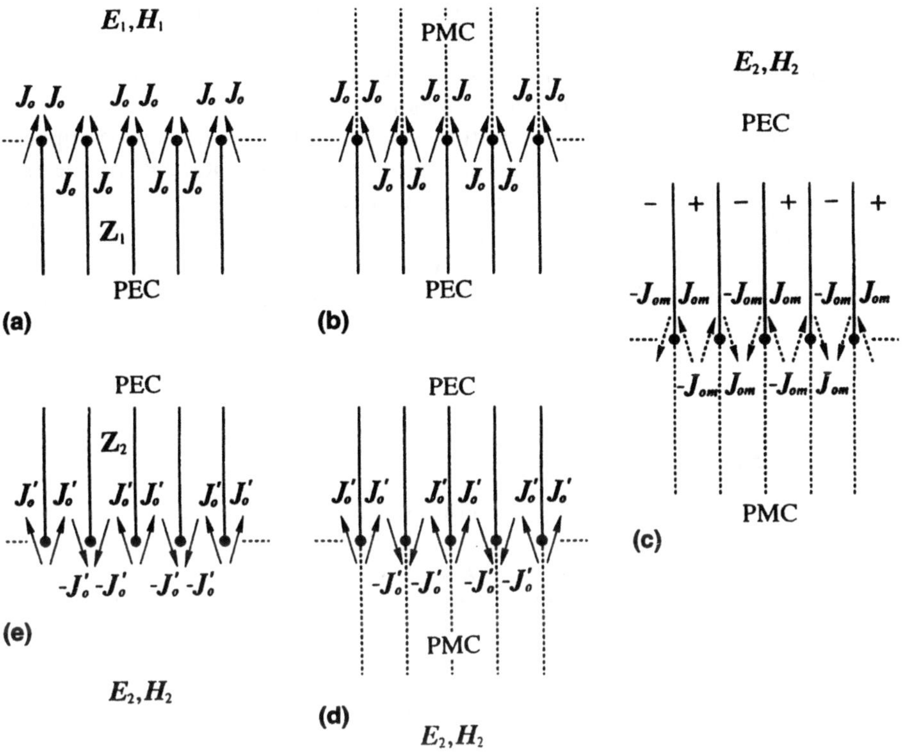

Fig. 7.6 Illustration of side-by-side periodically stacked pseudo-self-complementary structures:
(a)→(b) introduction of MPC sheets;
(b)→(c) transformation to the dual structure;
(c)→(d) equivalent transformation of source currents;
(d)→(e) elimination of MPC sheets.
(a) and (e) are complementary but not identical.

By comparing the structure (e) finally obtained with the original structure (a), it is found that their geometrical shapes are identical although the distributions of the source currents are different. This means that the configurations of the two structures in (a) and (e) are not identical as a whole, and hence both are not self-complementary as an antenna system, though all of the element antennas are themselves self-complementary.

Incidentally, it can easily be appreciated that the input impedances for all elements in (a) and (e) are respectively identical. Let their values be Z_1 and Z_2 for (a) and (e), respectively. Then, from the mutually complementary relationship between them we obtain

$$Z_1 Z_2 = (Z_0/2)^2 \tag{7.12}$$

However, no further analysis can be made for the structures in (a) and (e), since they are not self-complementary.

In order to avoid the difficulty explained above, an alternate periodic arrangement of a side-by-side stacked antenna system is proposed, where semi-infinite PEC sheets are inserted at the centre of every partial space between two adjacent element antennas, as illustrated in Fig. 7.7(a).

The same procedure of transformations as shown in Fig. 7.6 can be applied to this structure. In other words, introduction of complementary PMC sheets [(a) → (b)], transformation to the dual structure [(b) → (c)], equivalent transformation of source currents [(c) → (d)], and the elimination of PMC sheets [(d) → (e)] finally leads to the PEC structure (e). By comparing the original structure (a) with the structure (e) finally derived in Fig. 7.7, it will be

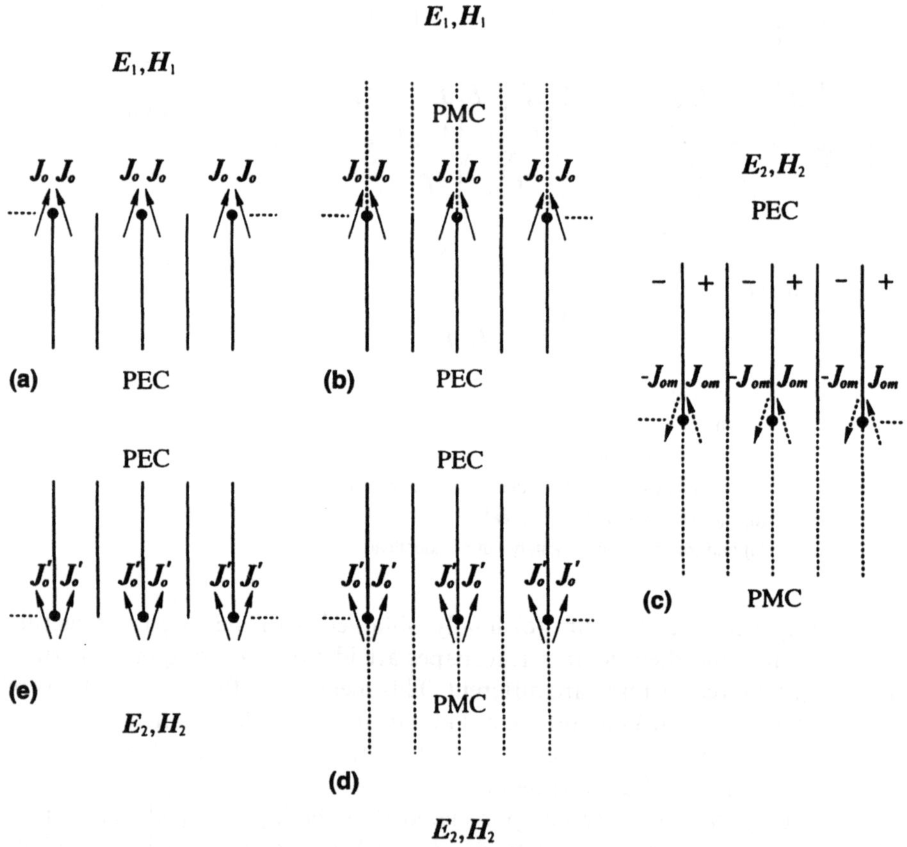

Fig. 7.7 Illustration of side-by-side periodically stacked self-complementary structure with semi-infinite periodical shields:

(a)→(b) introduction of MPC sheets;

(b)→(c) transformation of the dual structure;

(c)→(d) equivalent transformation of source currents;

(d)→(e) elimination of MPC sheets

(a) and (e) are complementary and identical.

found that their entire configurations are identical. In addition, they are mutually complementary owing to the procedures used in their derivation. This actually implies that the structures are self-complementary.

Accordingly, the input impedance Z_1 for each terminal in (a) and Z_2 for each terminal in (e) are equal, and they have a common value of Z. In addition, from the complementary relationship between (a) and (e) we have

$$Z_1 Z_2 = (Z_0/2)^2 \tag{7.13}$$

and consequently, we find

$$Z_1 = Z_2 = Z = Z_0/2 \tag{7.14}$$

This equation means that the input impedance at each terminal of Fig. 7.7(a) or (e) is always constant, independently of the source frequency and shape of the element antennas, including the effects of mutual coupling between the element antennas of this antenna system.

Thus, the stacked antenna in Fig. 7.7 has been proved to be self-complementary, but it includes semi-infinite PEC sheets where no radiation element is mounted. For this reason, another stacked antenna is proposed, for the purpose of avoiding the introduction of such useless elements.

The proposed structure is shown in Fig. 7.8(a), where all element antennas are self-complementary and all adjacent elements are mutually complementary. Hence it is an alternate periodic structure. This stacked antenna system can be treated by the same procedure used in the case of Fig. 7.7, and we obtain the structure given in Fig. 7.8(e). By comparing the structures in (a) and (e), we find that they are identical. Furthermore, they are evidently mutually complementary as a consequence of their derivation procedure.

Accordingly, the input impedance Z at each terminal in (a) and (e) is also given by

$$Z = Z_0/2 \tag{7.15}$$

Figure 7.9 illustrates an example of a side-by-side stacked self-complementary antenna more explicitly, since the structures in Figs 7.7 and 7.8 show only the principle of the stacking.

Incidentally, many other types of alternate periodic side-by-side stacked self-complementary antennas are conceivable as variations of the antennas described above.

7.5 Compound-stacked self-complementary antennas

By combining the principles of co-planar stacking and side-by-side stacking, compound-stacked self-complementary antennas can be derived as shown in Fig. 7.10.

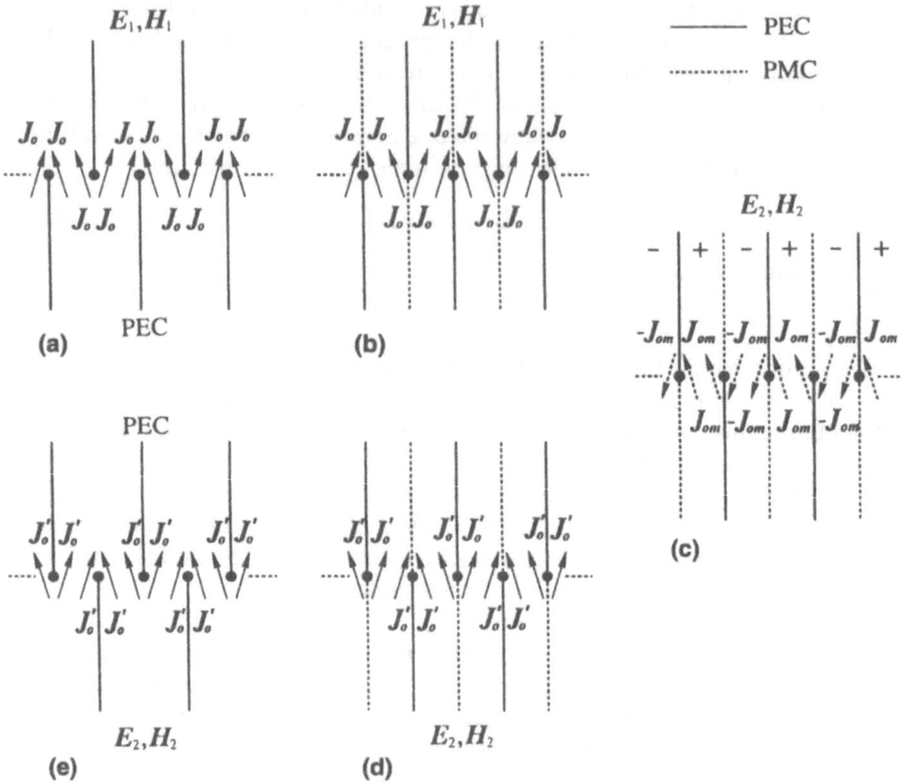

Fig. 7.8 Illustration of side-by-side alternately periodical stacked self-complementary structure:
(a)→(b) introduction of MPC sheets;
(b)→(c) transformation to the dual structure;
(c)→(d) equivalent transformation of source currents;
(d)→(e) elimination of MPC sheets
(a) and (e) are complementary and identical.

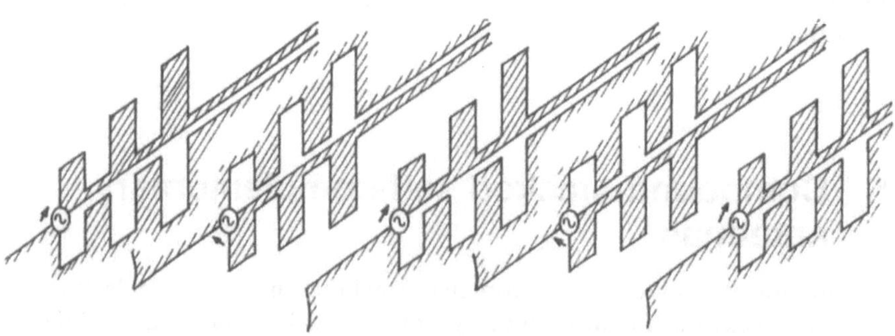

Fig. 7.9 An example of a side-by-side stacked self-complementary antenna ($Z \simeq 60\pi\ \Omega$ each).

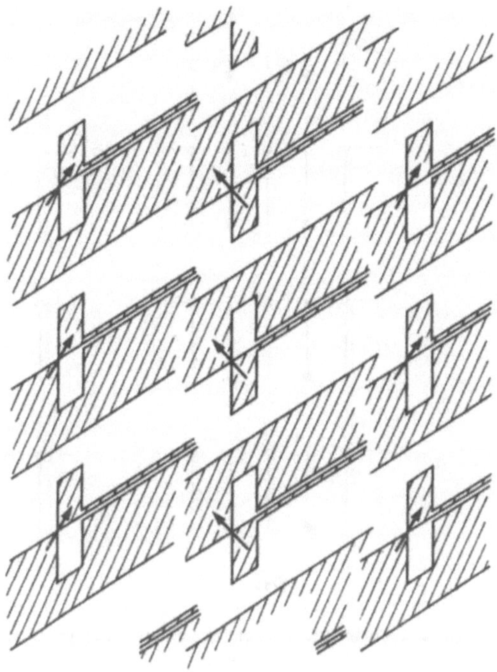

Fig. 7.10 Principle of the compound-stacked self-complementary antenna ($Z \simeq 60\pi\ \Omega$ each).

The input impedance at each terminal in this structure can easily be obtained as

$$Z = Z_0/2 \tag{7.16}$$

which is always constant, whatever the source frequency and shape of the structure.

The example mentioned above illustrates the principle for a typical case of two-dimensional stacking. However, many other variations of such stacking are conceivable, and some are discussed below.

For this purpose, greatly simplified symbols are used in Fig. 7.11. The open small circle stands for a radiation element with its exciting current as shown in Fig. 7.5(a), and the filled small circle stands for that shown in Fig. 7.5(c). Under such definitions of the symbols, Fig. 7.11(a) represents the structure in Fig. 7.10.

The structure symbolized by Fig. 7.11(b) represents one variation of the structure in (a). As understood from 7.11(b), one of two kinds of element structure is shifted in the vertical direction by a half-period. For such a structure, it is not difficult to show that the shape of the complementary structure is identical to the original one, and that the structure is self-complementary. In this case, however, if the shift distance is not equal to the half-period, then the complementary structure is not identical to the original one, and it is not self-complementary.

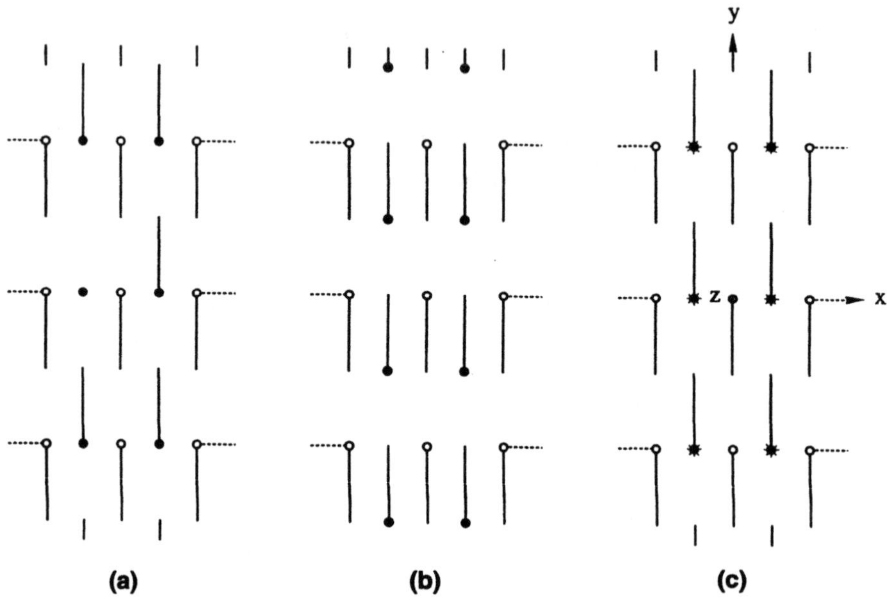

Fig. 7.11 Various types of compound-stacked self-complementary antennas sketched in greatly simplified symbols ($Z \simeq 60\pi\ \Omega$ each):
(a) same as Fig. 7.10;
(b) elements alternately shifted in the y-direction by half-period;
(c) elements alternately shifted in the z-direction.

Another variation is shown in Fig. 7.11(c), where the element structures symbolized by a 'star' are shifted in the z-direction uniformly with reference to the structure in (a). This structure can be proved to be self-complementary as a whole, and the input impedance at each feed point is also given by (7.16). Other variations of compound stacking are conceivable, but no further description will be given here.

8 • GENERAL CONSIDERATIONS ABOUT APPROXIMATIONS AND MODIFICATIONS OF SELF-COMPLEMENTARY ANTENNAS

8.1 Approximations and modifications for practical purposes

According to the theory of the self-complementary antenna (SCA), the basic structures of this type of antenna consist of infinitely extended planar sheets of perfect electric conductors. In addition, in the case of stacked antennas, an infinite number of element antennas are arranged periodically. Therefore, the theoretically obtained self-complementary antenna can only be approximated by limited size of structure and limited number of element antennas, when this principle is applied to extremely broad-band practical antennas.

Furthermore, for the purpose of avoiding mechanical structural weakness in this type of antenna, most of the conducting sheet structures must be approximated by electrically equivalent and mechanically firm conducting structures.

Additionally, since the self-complementary property of an antenna prescribes nothing about the broad-band nature of its radiation properties, various information about the radiation characteristics of other existing antennas must be introduced, for the purpose of developing practical broad-band antennas. In this connection, various results of experimental studies show that some modifications to the shapes of antennas are very effective in improving their radiation characteristics, without causing much deterioration to the frequency band-width of their input impedances.

In the following sections, the approximations and modifications mentioned above for self-complementary antennas will be discussed, the aim being the practical application of this type of antenna in extremely broad-band antennas.

8.2 Approximation by truncation

The most indispensable approximation to theoretically derived self-complementary antennas is truncation of their infinitely extended conducting sheets, for the cases of both the single antenna and the stacked antenna. In the course of such approximations, if the current distributions on the planar sheets are disturbed by the truncation, then there will be some harmful effects at the feed point as a result of reflected currents from the truncated ends. Such truncation effects cause deterioration of the constant-impedance property assured by the theory.

For this reason, the reduction of the truncation effects is very important, and the truncation must be made at a location where the current distribution density is sufficiently attenuated. However, it is very difficult to estimate the precise current distributions, and hence experimental "hard" studies are mainly needed for the purpose of reducing the truncation effects.

In this connection, an instructive fact has been found from the results of experimental studies performed by the author's group: namely, the reduction of truncation effects in the "teeth-type" array structure is satisfactory in general [see, for example, 8.1–8.3]. This property can be explained as follows.

The "teeth-type" array structure enhances radiation, and attenuation of the current distribution from the feed point is very rapid. This reduces the reflected current from the truncated end, and reduction of the truncation effects becomes satisfactory. The equally spaced tapered array [8.1] and the log-periodic tapered array [1.12] are typical examples of such structures. In these cases, however, the largest dimensions of the truncated antennas determine the low-frequency limit for the constant-impedance property of antennas in practice.

8.3 Approximation by replacement with conducting rods

In general, a narrow strip of conducting sheet in an antenna structure can be replaced by an equivalent conducting rod, under the condition that the width of the strip is sufficiently small compared with the wavelength. Such a replacement is very effective for improving the mechanical strength of sheet structure in self-complementary antennas.

Now, according to the theory of linear antennas, an equivalent radius ρ_e of a narrow strip of conducting sheet, as shown in Fig. 8.1(a), is given [1.18] by

$$\rho_e = W/4 \tag{8.1}$$

where W is the width of the strip, and "equivalent radius" means the radius of the circular cross-section of a conducting rod that has the same characteristics as a linear antenna.

On the other hand, the equivalent radius of two parallel rods is given [1.18] by

$$\rho_e \simeq \sqrt{dr} \tag{8.2}$$

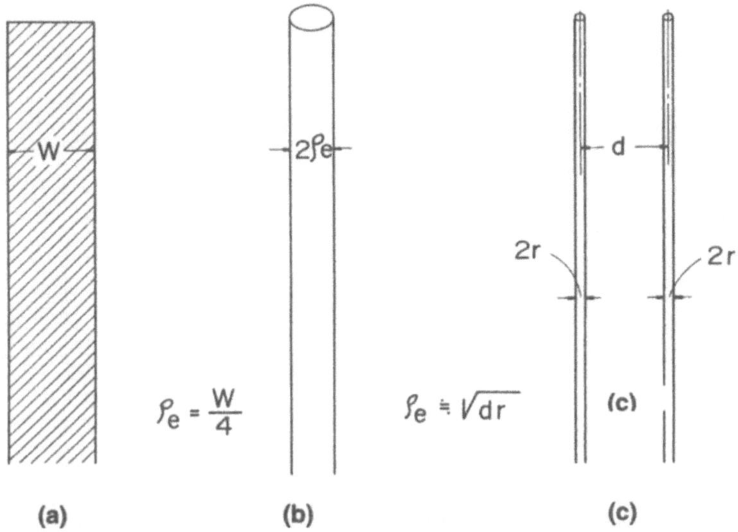

Fig. 8.1 A narrow strip of conducting sheet and its approximation by equivalent conducting rods.

where r is the radius of the circular cross-sections of the rods and d is the separation distance between the centre-lines of the rods, as shown in Fig. 8.1(c). Therefore, a narrow strip of conducting sheet can also be approximated by two parallel conducting rods using relationship (8.2), though there is freedom in choosing the dimension of either one or the other so as to give a particular value to ρ_e. For example, the following two particular cases are practically interesting.

Case (1): When $d = W$ is assumed, then from (8.1) and (8.2) we find

$$r \simeq W/16 \tag{8.3}$$

Case (2): When $d = W/2$ is assumed, then, also from (8.1) and (8.2) we obtain

$$r \simeq W/8 \tag{8.4}$$

These results for the two cases show that a narrow conducting strip sheet as shown in Fig. 8.2(a) can be replaced equivalently with a trapezoidal wire structure either as in Fig. 8.2(b) or in Fig. 8.2(c). The structure in Fig. 8.2(b) corresponds to Case (1), where the centre-line of the wire coincides with the outer edge of the sheet [8.4]. In contrast, Fig. 8.2(c) corresponds to Case (2), where the gap width between the two rods is equal to the diameter of the rod.

Also, a fan-shaped conducting sheet with narrow apex angle θ, as shown in Fig. 8.3(a), can be approximated by a circular cone as shown in Fig. 8.3(b). The equivalent cone angle ψ can be given as

$$\psi \simeq \theta/4 \tag{8.5}$$

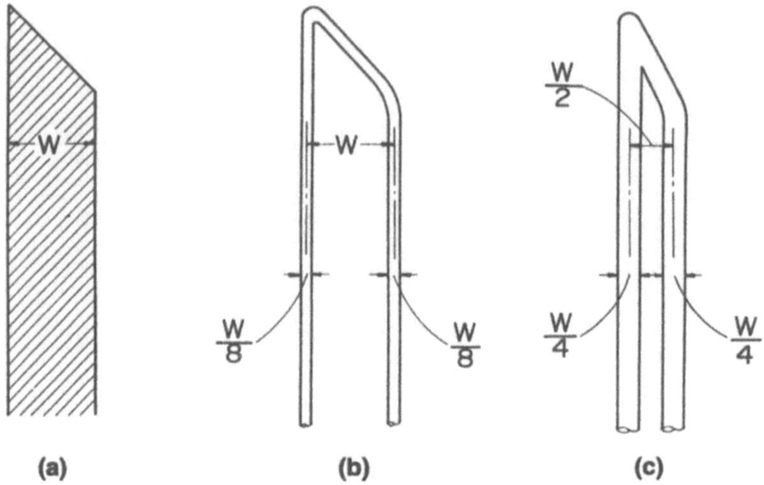

Fig. 8.2 Examples of approximation by conducting rods: (a) is original structure, (b) and (c) are approximations.

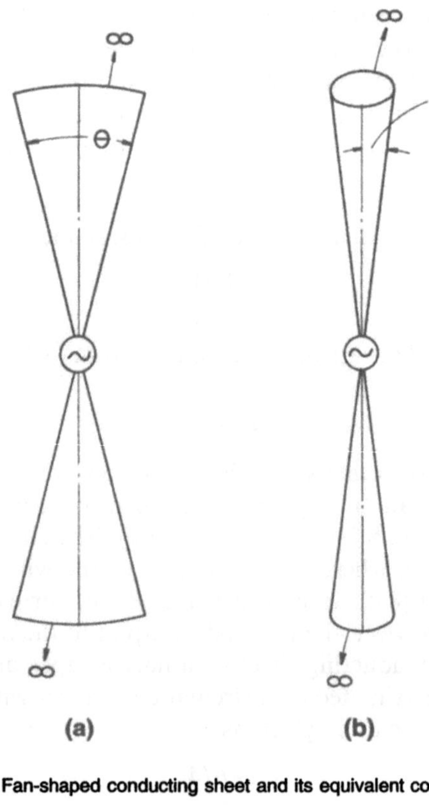

Fig. 8.3 Fan-shaped conducting sheet and its equivalent conducting cone.

corresponding to (8.1). Incidentally, the characteristic impedance Z of the conical transmission line in Fig. 8.3(b) for transverse electromagnetic mode (TEM) is given [8.5] by

$$Z \simeq 120 \ln \cot(\psi/2) \simeq 120 \ln \cot(\theta/8) \qquad [\Omega] \tag{8.6}$$

Fortunately, the equivalent radii of the rod structures and the equivalent cone angle discussed above are independent of the source frequency provided the radii remain sufficiently small compared with the wavelength. For this reason, substitution with the rods is an advantageous approximation for the sheet structures of self-complementary antennas.

8.4 Modification by deformation

The broad-band nature of the input impedances of self-complementary antennas is well preserved, even when their planar structures are deformed by bending or folding, although the values of their input impedance are transformed to other values. On the other hand, the radiation patterns of antennas that consist of planar structures can be changed considerably by deformations such as bending or folding. By making use of such effects of deformation, the radiation characteristics of self-complementary antennas can be improved in practice, without causing much deterioration in the broad-band nature of the input impedance.

Although the introduction of a ground plane may not be considered as a deformation, it is evidently a type of modification. In general, it is often effective to use the ground plane for the purpose of improving the radiation characteristics of antennas. In principle, however, the introduction of the ground plane disturbs the self-complementary configuration, but sometimes the broad-band characteristics of the input impedance are preserved to some extent [8.6, 8.7].

8.5 Modification by partial excision

During the studies of three-dimensional self-complementary antennas, it was found that the lower half of the vertical portion shown in Fig. 6.7 can be excised without losing the constant-impedance property, although the value of the impedance transforms from $Z_0/4$ to $Z_0/\sqrt{2}$ [8.8–8.10]. This type of modified self-complementary antenna (MSCA) will be discussed later in Chapter 11.

8.6 An example of transformation from a self-complementary sheet structure to the conducting rod structure

The log-periodic antenna and the log-periodic dipole array have excellent characteristics as extremely broad-band antennas. However, the broad-band characteristics of these antennas have their origin not in the log-periodic shape, but rather in the aspects of the shape that are derived from the self-complementary antenna, as explained in Chapter 1, section 1.1.

The above-mentioned broad-band property of log-periodic structures is a natural consequence for derivatives of self-complementary structures. In order to make this reason more understandable, the detailed procedures of the transformation will be explained below [8.10], as an example of the modification and approximation described in the preceding sections.

For this purpose, the equally spaced self-complementary antenna shown in Fig. 4.3(a) is considered as an original structure. The oblique view of the same structure is shown in Fig. 8.4, where only the principal portion of the structure near to the feed point is shown, for simplicity. It is easily understood that the radiation from this antenna is bi-directional. However, for obtaining uni-directional radiation, a modification to the conical shape as shown in Fig. 8.5 is conceivable, and has been successfully performed by investigators in the United States [1.12, 1.13, 8.4].

The next step of transformation is modification of the shapes of the sheet strips, that is, deformation from circular arcs to the straight strips as shown in Fig. 8.6.

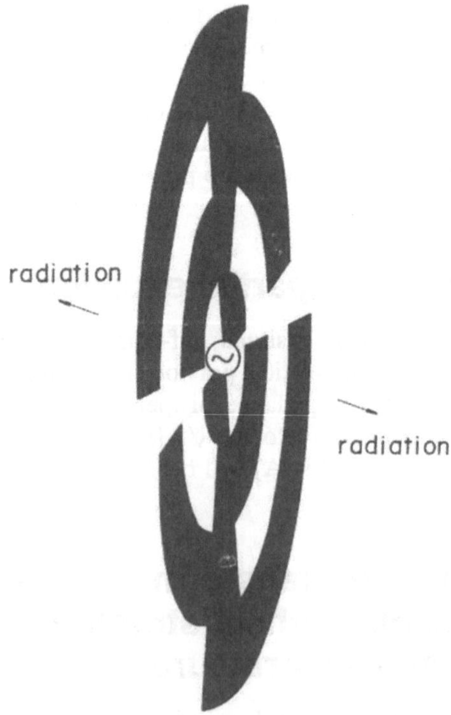

Fig. 8.4 Equally spaced SCA in finite size.

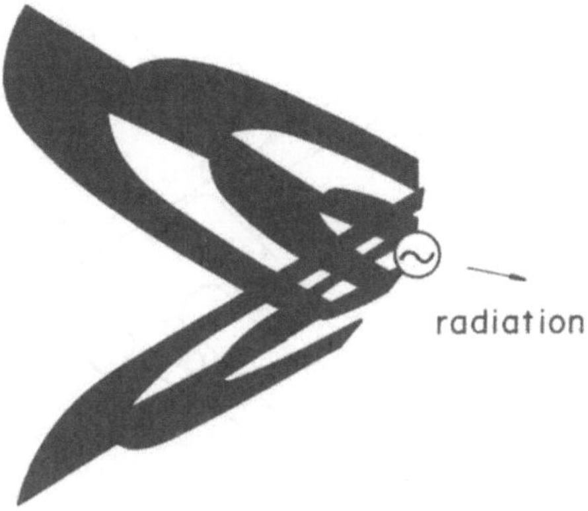

Fig. 8.5 MSCA on an angularly bent plane.

Fig. 8.6 MSCA in conical shape.

This structure can also be derived from the self-complementary antenna shown in Fig. 4.3(b), which has straight branch-strips, instead of the circular-arc branch-strips shown in Fig. 4.3(a). Hence, the straight strips in Fig. 8.6 can be approximated by conducting rods with an equivalent circular cross-section, as illustrated in Fig. 8.7.

Also, another approximation with two parallel rods can be made by applying the principle shown in Fig. 8.2, and this leads to the structure shown in Fig. 8.8.

This last structure corresponds to the so-called log-periodic antenna if the original antenna is replaced by the log-periodic self-complementary antenna. Incidentally, an example of such a modified self-complementary antenna in practical use is shown in Fig. 8.9, which is widely known under the name of "log-periodic antenna".

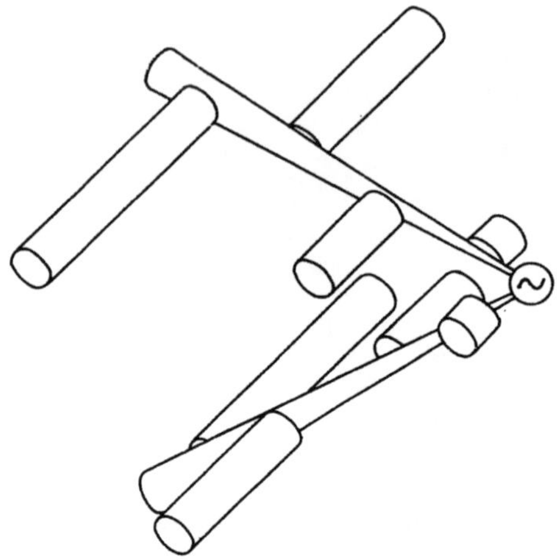

Fig. 8.7 MSCA formed with conducting rods.

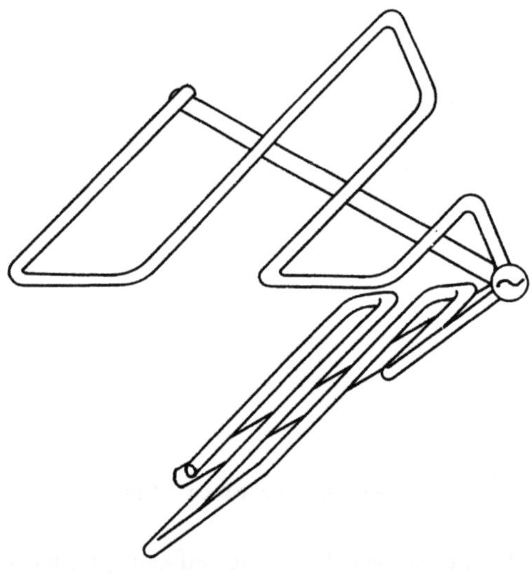

Fig. 8.8 MSCA formed with bent conducting rods.

As understood from the procedure of transformation described above, the antenna shown in Fig. 8.8 is evidently a derivative of the self-complementary antenna, and the broad-band property of such a type of antenna comes from the self-complementary nature of the original antenna. Accordingly, if one of the two wings of the antenna in Fig. 8.8 is arranged upside-down, then the broad-band characteristics cannot be assured in such a structure, even when the original structure is log-periodic. In fact, it was confirmed by experiment that an antenna with log-periodic structure, but arranged improperly as mentioned above, has distinctly varying log-periodic input impedance for the

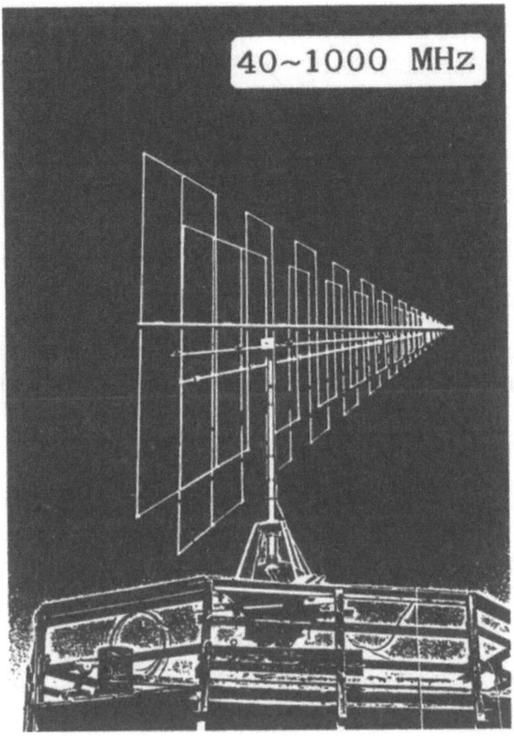

Fig. 8.9 An example of MSCA in practical use (known as log-periodic antenna, by courtesy of Denki Kogyo Co., Ltd., Tokyo).

source frequency [1.6]. Hence, it does not have frequency-independent characteristics as mentioned in Chapter 1, section 1.1, and also as explained in Chapter 9, section 9.4.

Finally, a rather bold modification as illustrated in Fig. 8.10 is considered. The two-wing conical structure in Fig. 8.7 is completely folded, to form a coplanar dipole array. In this structure, it should be noted that the directions of the excitation from the parallel-conductor line to the dipoles are alternately interchanged. In other words, this dipole array is in transposed excitation. For this reason, it can be understood that the transposed excitation is a natural consequence of its derivation procedure from the self-complementary structure. Therefore, if the original antenna is assumed to be a self-complementary log-periodic structure, then the final structure becomes a so-called log-periodic dipole array. This means that the log-periodic dipole array excited without transposition is not a derivative of the self-complementary antenna, and the good broad-band characteristics cannot be assured for such a structure.

In conclusion, the so-called log-periodic antenna and the log-periodic dipole array are derivatives of the self-complementary antenna, and the origin of their broad-band characteristics is in the self-complementary nature of their original structure, and not in their log-periodic shape.

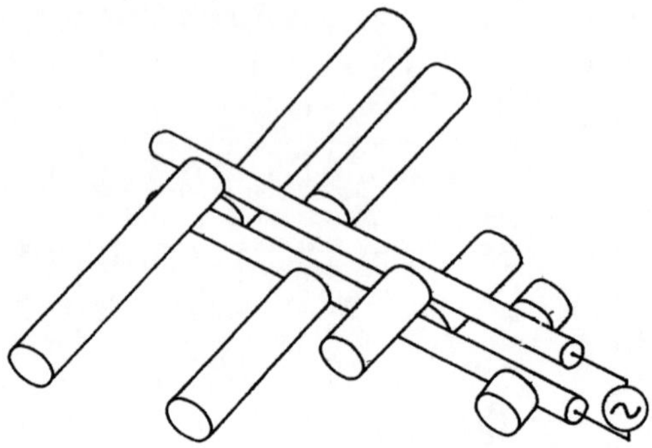

Fig. 8.10 MSCA formed with conducting rods on a folded plane.

9 • DEVELOPMENTAL STUDIES OF ROTATIONALLY SYMMETRIC SELF-COMPLEMENTARY ANTENNAS

9.1 Alternate-leaves type self-complementary antenna

In the process of developing extremely broad-band practical antennas on the basis of the theoretical principle of self-complementary antennas, it is always necessary to make some additional experimental studies as mentioned in Chapters 1 and 8. For this reason, any basic information related to the measured characteristics of practical self-complementary antennas (SCA) will be helpful for their developmental investigations in general. Therefore, it is worthwhile obtaining various experimental data to the greatest extent possible.

To this end, the characteristics of equally spaced alternate-leaves type SCA have been studied experimentally from various points of view, as a typical example of rotationally symmetric SCA in finite size. The geometry of this type of antenna is shown in Fig. 9.1, where the various dimensions for Antenna A are indicated [9.1].

During experimental studies of this type of antenna, four different models were prepared, and these were designated respectively as Antenna A, Antenna B, Antenna C and Antenna D. The dimensions of various parts of the structures for these four antennas are listed in Table 9.1. Here, it must be noted that Antenna C is not self-complementary, since it was prepared just for comparison purposes.

The measured input impedance of Antenna A for frequencies from about 0.5 to 2.3 GHz is plotted as a Smith chart in Fig. 9.2, where the values of impedance are normalized by 50 Ω corresponding to the impedances of the feeder and the measuring equipment. This chart shows that the variation of the input impedance is extremely small for frequencies higher than 0.61 GHz, where the length L of the longest leaf of the antenna is equal to a quarter-wavelength for that frequency. Let such a frequency be denoted by f_L hereafter. Incidentally, at this frequency the total length of the periphery of this antenna is approximately equal to the corresponding wavelength.

It is found from this experiment that the constant-impedance property of this type of self-complementary antenna has been well proven for frequencies

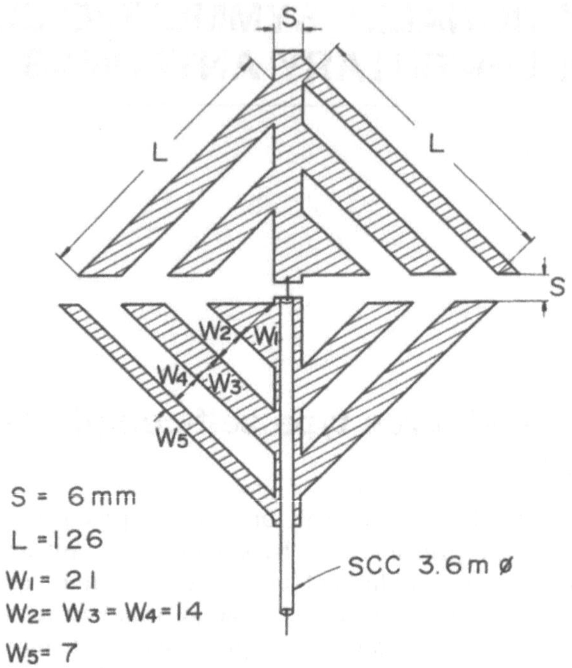

Fig. 9.1 Alternate-leaves type SCA.

Table 9.1 Dimensions of alternate-leaves type SCAs in mm (see Fig. 9.1)

	Antenna A	Antenna B	Antenna C*	Antenna D
Numbers of leaves	5 × 2	9 × 2	5 × 2	12 × 2
S	6	6	6	6
L	126	126	126 (112)	322
W_1	21	14	14 (21)	21
W_2	14	7	21 (7)	14
W_3	14	7	7 (21)	14
W_4	14	7	21 (7)	14
W_5	7	7	7	14
W_6		7		14
W_7		7		14
W_8		7		14
W_9		7		14
W_{10}				14
W_{11}				14
W_{12}				14

*Antenna C is not self-complementary.

higher than f_L, the quarter-wave frequency for the longest leaf of this antenna. In other words, the effects of truncation of finite size are sufficiently small for such frequencies, notwithstanding that the constant-impedance property is theoretically true only for infinitely extended self-complementary structures. This means that reduction of truncation effects is very pronounced for an equally spaced teeth-type structure, and such a benefit is not limited just to log-periodic structures.

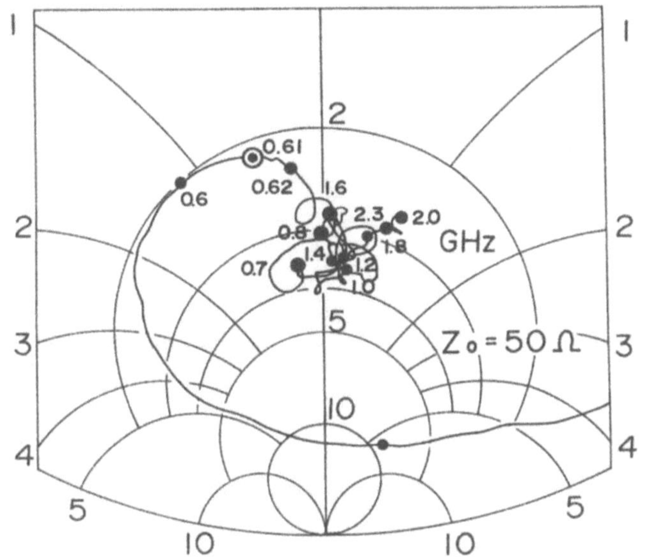

Fig. 9.2 Measured input impedance of Antenna A (0.5–2.3 GHz), ($Z_0 = 50\ \Omega$).

Next, the measurements of the frequency characteristics of the input impedance were also made for Antenna B. Here, the geometrical outer size of this antenna is exactly equal to that of Antenna A, but the incision of the teeth-type leaves for Antenna B is twice as fine as that for Antenna A, as understood from the dimensions listed in Table 9.1. It was found from the results of the measurements that the frequency characteristics of the input impedance of Antenna B are almost in the same form as those in Fig. 9.2 obtained for Antenna A, and hence the impedance chart is not shown here.

In order to examine the frequency characteristics of the input impedance for other structures of this type, measurements have also been made for Antenna C, which is not self-complementary. This antenna is derived as a modification of Antenna A, by making the width of its alternate-leaves narrower to become one-half of that for Antenna A, without changing the original skeletal structure, as understood from Table 9.1. Accordingly, the gaps between the adjacent leaves in this antenna become larger than the widths of the leaves, and hence the self-complementary condition is not satisfied for Antenna C.

The measured impedance for Antenna C is plotted in Fig. 9.3. This Smith chart shows that the concentration of values of input impedance to $60\pi\ \Omega$ has deteriorated distinctly when compared with that of Fig. 9.2, as expected from existing theory.

In order to obtain further experimental data, input impedance measurements were also made for Antenna D, which is the largest model among the prepared four antennas. The results of the measurements are plotted in the Smith chart shown in Fig. 9.4.

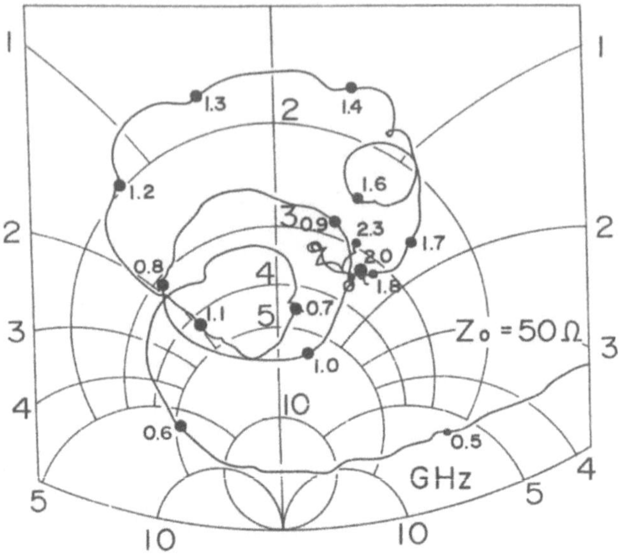

Fig. 9.3 Measured input impedance of Antenna C (0.5–2.3 GHz), ($Z_0 = 50\ \Omega$).

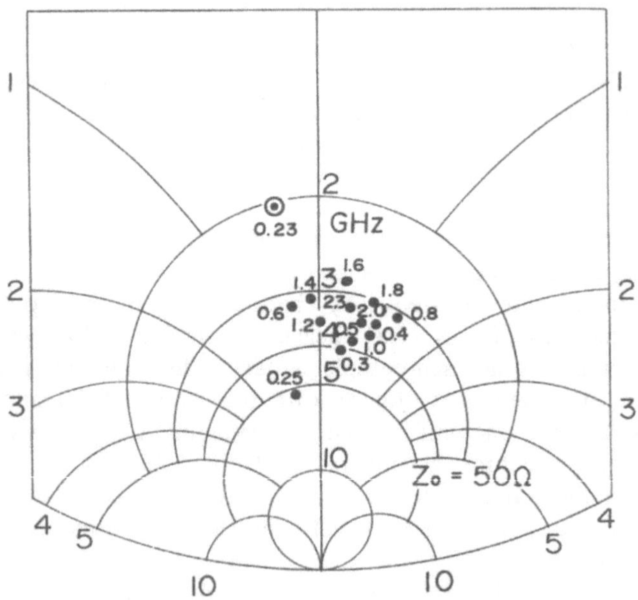

Fig. 9.4 Measured input impedance of Antenna D (0.23–2.3 GHz), ($Z_0 = 50\ \Omega$).

By comparing this figure with Fig. 9.2, a distinct difference can be seen between the behaviour of the impedances for the frequencies below 0.6 GHz. In other words, the concentration of impedance values to 188 Ω for Antenna D occurs over the frequency range from 0.23 to 2.3 GHz, while the corresponding frequency range for Antenna B is from 0.61 to 2.3 GHz. This means that the band width of the input impedance for Antenna D is much broader than that for Antenna B, as a natural consequence of the difference in antenna size. Incidentally, the quarter-wave frequency for the longest leaf is given as f_L = 0.23 GHz for Antenna D, while the frequency f_L is equal to 0.61 GHz for Antenna B.

The discussions above have been limited to the impedance properties of the alternate-leaves type antenna. However, for the purpose of developing practical broad-band antennas, it is also indispensable to secure the broad-band property of their radiation characteristics as well as their constant-impedance property. For this reason, the radiation patterns of alternate-leaves type SCA have also been studied experimentally.

In this experiment, the rectangular coordinates, x, y, z, and the angular coordinates, θ, ϕ, are assumed as indicated in Fig. 9.5.

From the measured radiation patterns of Antenna A in the yz-plane and the xz-plane for various frequencies from 0.6 to 1.9 GHz, it is found that the patterns remain almost unchanged over this frequency range. Some examples of the measured patterns for 0.8 and 1.6 GHz are shown in Figs 9.6(a) and (b). In these measurements, it should be noted that the major electric field of radiation is in the y-direction.

The measurements for Antenna B were also made in the same way, and it was found that the radiation patterns were almost identical to those in Fig. 9.6 obtained for Antenna A, except that the cross-polarization components that exist for Antenna A in the range of higher frequencies are suppressed for Antenna B. Such an improvement in polarization of the radiated field can be

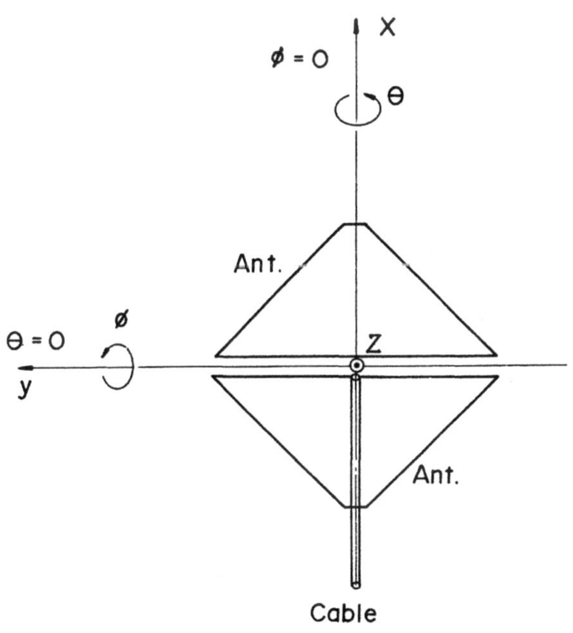

Fig. 9.5 Coordinates for the antenna.

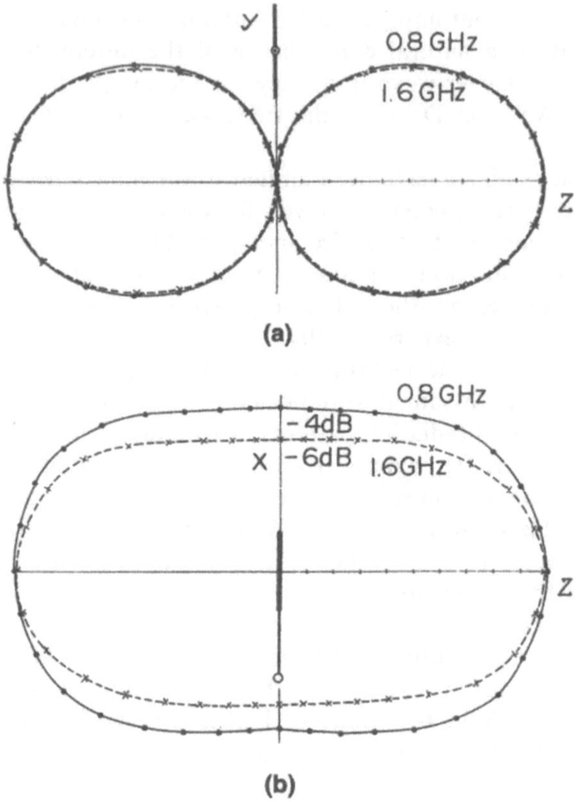

Fig. 9.6 Measured field patterns for Antenna A: (a) *yz*-plane, (b) *xz*-plane.

attributed to the finer incision of the alternate-leaves for Antenna B. For this reason, it can be understood from the results of these experiments that the finer structure is desirable for shorter leaves in the range of higher frequencies.

Further measurements were made for Antenna D in the broader frequency range of 0.2–1.9 GHz, and almost the same results as in the case of Fig. 9.6 were obtained, except for small irregularities in the radiation patterns for frequencies higher than 1.4 GHz. As mentioned above, such a difficulty could be avoided by designing finer structures for shorter leaves, instead of adopting uniform width for all leaves. One example of such a structure is the alternate-leaves type log-periodic SCA, though the periodicity is not a necessary condition.

9.2 Approximation and modification of alternate-leaves type self-complementary antennas

A simple approximation for the structure of the alternate-leaves type SCA is the replacement of its leaves with equivalent conducting rods, as explained in

Chapter 8. Such an example of approximated structure was fabricated by applying relationship (8.1) to the structure of Antenna A [9.1]. The results of impedance measurements for this antenna showed that the same level of constant-impedance property as that for Antenna A is also well preserved for this approximated antenna, as expected from the theory.

Incidentally, the truncation of the infinitely extended theoretical self-complementary structure to a certain finite size, is an indispensable approximation in the process of developing a practical broad-band antenna. However, the results of impedance measurements for Antennas A, B and D show that the truncation effects are extremely small for this kind of teeth-type array structure, except for frequencies lower than f_L. This means that the truncation of structures designed for lower frequencies must be made at larger dimensions corresponding to the wavelength.

Next, an example of modification of Antenna A was studied by deforming this planar structure into a bent structure. That is, the two wings of this antenna were on different surfaces of a wedge-shaped bent plane. A side view of such a deformed antenna with slanted wings is shown in Fig. 9.7, where the angle ψ becomes 180° for Antenna A, the original structure.

The results of impedance measurements for varying angles of ψ showed that the constant-impedance property is well preserved even when the angle ψ is decreased to 30°, though values of the input impedance are shifted a little lower as angle ψ decreases.

For such modifications of Antenna A, variations of the radiation patterns were also studied. Results of the measurements showed that the bi-directional radiation pattern for $\psi = 180°$ transforms gradually to a uni-directional pattern for decreasing angles of ψ. It was also found from experiments that the front-to-back ratio of the radiation is 7–13 dB for $\psi = 90°$ and 8–16 dB for $\psi = 60°$, for frequencies higher than 1.0 GHz.

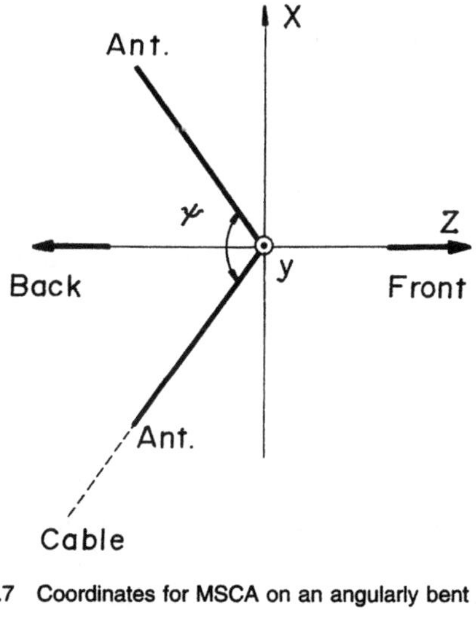

Fig. 9.7 Coordinates for MSCA on an angularly bent plane.

The results described above show that the modified alternate-leaves type SCA illustrated in Fig. 9.7 can be used as a uni-directional broad-band antenna.

In addition, another modification of Antenna A was studied experimentally, where the original shape of the square structure was deformed into a rhombic structure. In that deformation, the dimensions in the y-direction remain unchanged but the dimensions in the x-direction are proportionally reduced. In such a deformation, the flare angle of alternate leaves is increased from 90° to 120°. Here, a 90° flare angle is the condition for self-complementary structure, and hence the structure with 120° is not self-complementary. The results of impedance measurements on this deformed structure showed that the impedance values normalized to 188 Ω in the Smith chart occur approximately around or in the loci of voltage standing-wave ratio (VSWR) $\simeq$ 4. This means that the constant-impedance property is not preserved in such a modification, though a trace of the original property can still be found.

9.3 Modified four-terminal self-complementary antenna on conical surface

As an example of a developmental study of the self-complementary antenna (SCA) for practical use, a conically deformed four-terminal modified SCA for circularly polarized waves will be discussed [9.2]. The objective of this investigation was to develop a broad-band uni-directional practical antenna for circularly polarized waves, by making use of the principle of the SCA. For that purpose, three kinds of model antenna were fabricated. As the basic structure of these antennas, a proportionally spaced teeth-type array, or log-periodic structure, as shown in Fig. 9.8 was adopted.

Cones 1 and 2 are circular while Cone 3 is square pyramidal. All the cone angles are $2\theta = 48°$, and the lengths of the antenna wings are 55 cm, as shown in Figs 9.9(a), (b) and (c).

However, the dimensions in the horizontal direction are correctly reduced for all three models. For Cone 1, the horizontal lengths of the conducting sheets are equal to those of the space in every circular cross-section, or its dimensions in the horizontal direction are reduced by a factor of sin 24° compared with those in Fig. 9.8. In contrast, the horizontal lengths of the conducting sheets for Cone 2 are the maximum possible, in order that the dimensions of the gaps between the conductors in every horizontal cross-section are smaller than those for the conductors. Also, for Cone 3, the lengths of the conductors in horizontal cross-sections are equal to those for Cone 2 in cross-sections at the same height.

The measured input impedances between two opposite terminals are plotted as Smith charts, normalized to 50 Ω, in Figs 9.10(a), (b) and (c).

The filled circles show the values when the other two terminals are short circuited, while the open circles or squares show those when the terminals

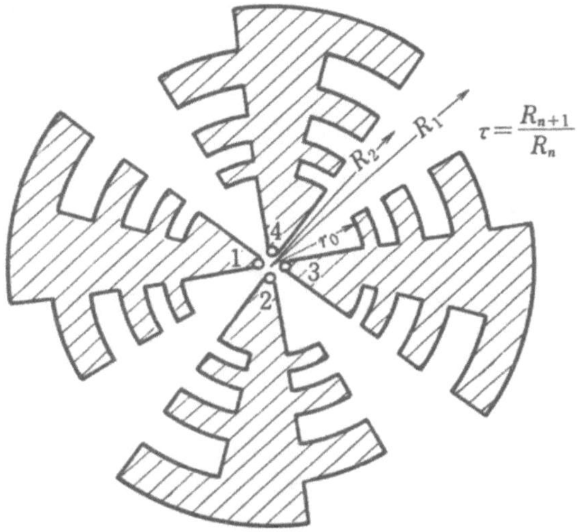

Fig. 9.8 Proportionally spaced SCA.

are opened. The theoretical impedance between two opposite terminals for the structure in Fig. 9.8 is $60\sqrt{2}\pi\ \Omega \simeq 266.4\ \Omega$. For Cone 1, the values of the input impedance on the chart are approximately around this value, in spite of the fact that the actual dimensions of the conducting sheets are reduced in the transverse direction by the factor of sin 24°. This fact suggests a very important property which is useful when a planar SCA is modified by deformation into a conical structure. Also, it is found experimentally that the input impedance between two opposite terminals is independent of the other two terminals, as expected from the theory. On the other hand, the concentration of impedance values on the chart is somewhat worse for Cone 2. In addition, the mean value of the impedance for the same cone becomes a little lower. For Cone 3, the concentration of impedance values on the chart is satisfactory, though the mean value becomes a little lower than $60\sqrt{2}\pi\ \Omega$.

Furthermore, various measurements of the radiation characteristics of these three antennas have also been made for turnstile excitation, but the results for Cone 2 will be omitted here, because the uni-directional property is a little worse than those of other models. Figures 9.11(a) and (b) show the field patterns for Cones 1 and 3.

Figure 9.12 shows the variation of the beam width with respect to the angle ϕ, while Fig. 9.13 shows the frequency dependence of the beam width in the plane $\phi = 0°$, and Fig. 9.14 shows the axial ratios of the polarization, also for Cones 1 and 3.

The various experimental results described above will provide very useful data for developmental studies for the uni-directional broad-band antennas, especially where the technique of deforming the planar SCA into a modified SCA of conical shape is to be applied.

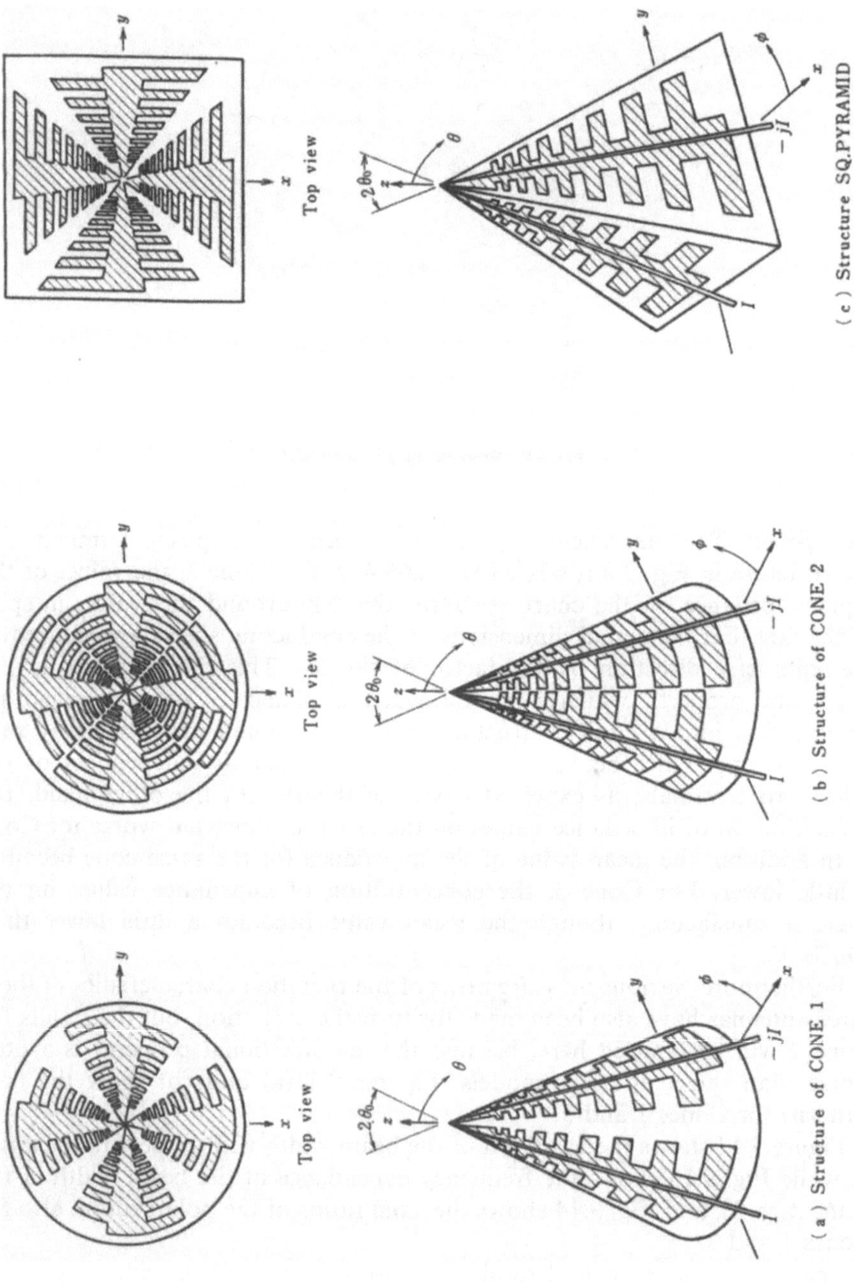

Fig. 9.9 MSCA on conical surface: (a) Cone 1, (b) Cone 2, (c) Cone 3.

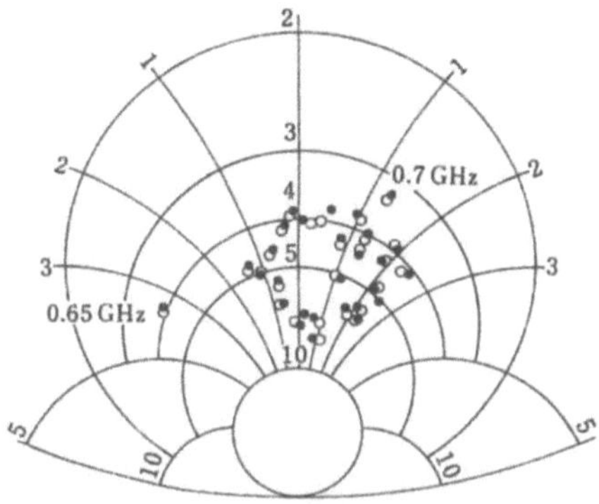

(a) For CONE 1 $f=0.65\sim2$ GHz (50 M step)

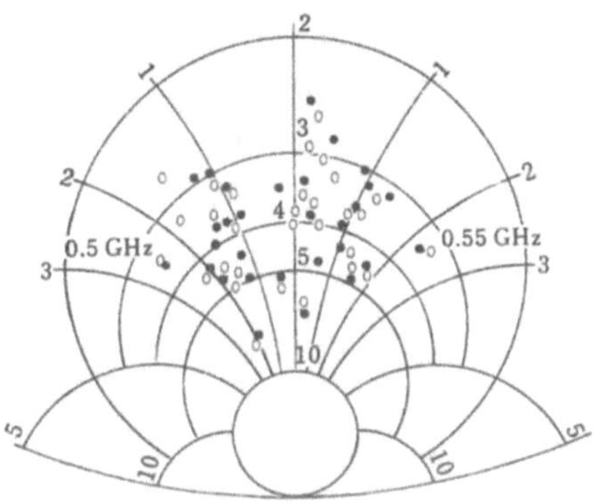

(b) For CONE 2 $f=0.5\sim2$ GHz (50 M step)

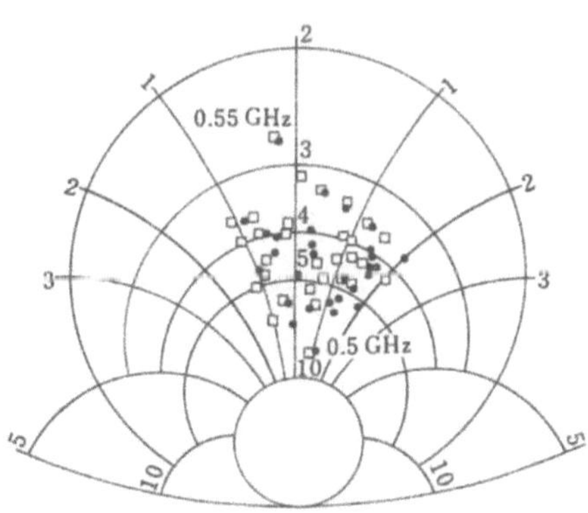

(c) For SQ. PYRAMID $f=0.5\sim2$ GHz (50 M step)

o , o , □ : opened : shorted

Fig. 9.10 Measured input impedances for MSCA on conical surface ($Z_0 = 50\ \Omega$): (a) Cone 1, (b) Cone 2, (c) Cone 3.

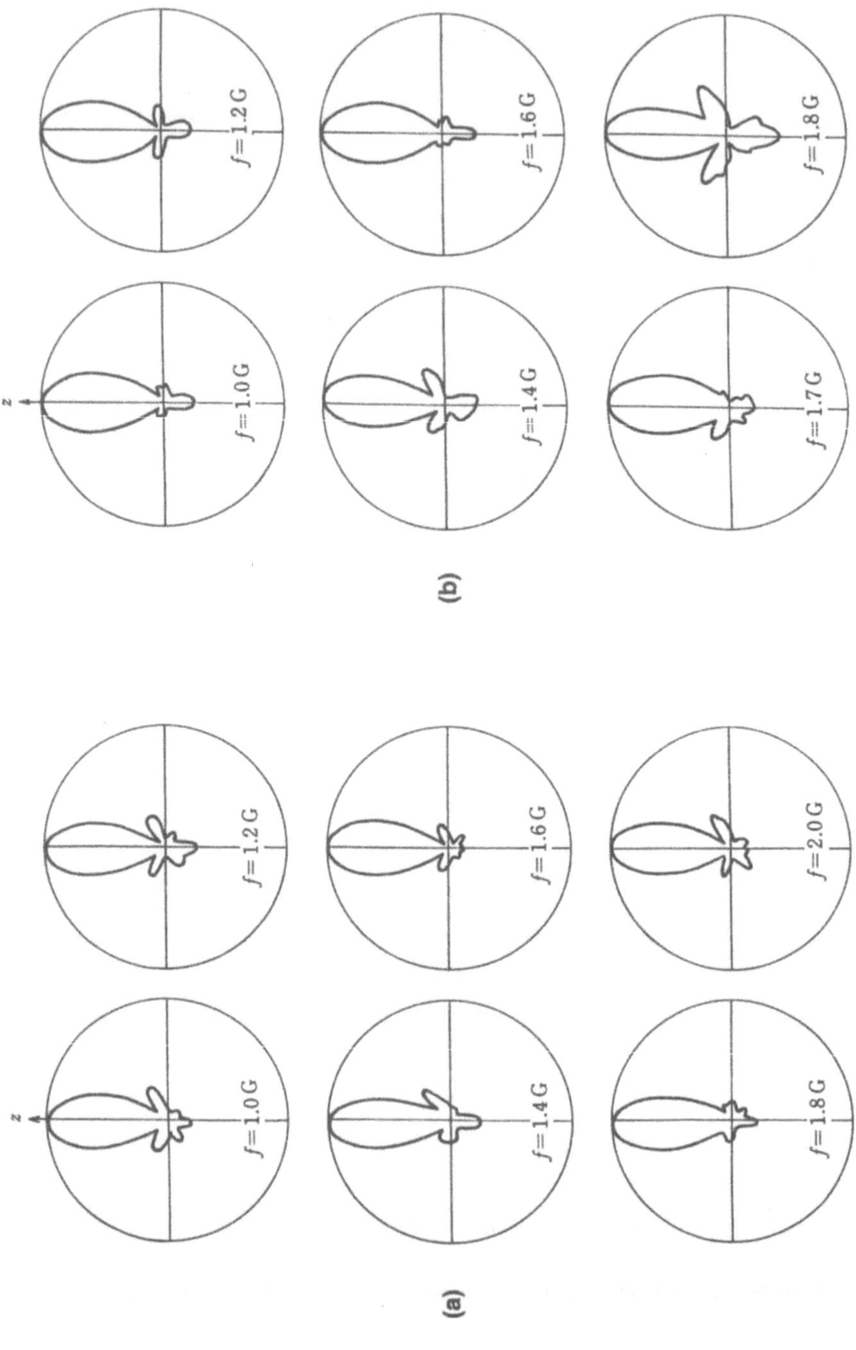

Fig. 9.11 Measured field patterns E_θ: (a) Cone 1, (b) Cone 3.

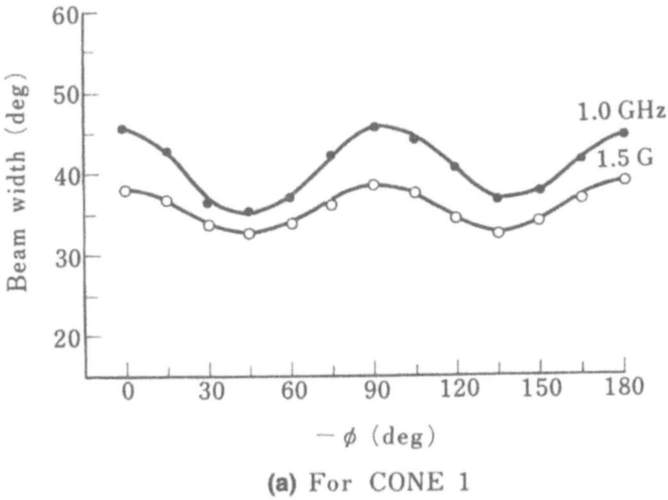

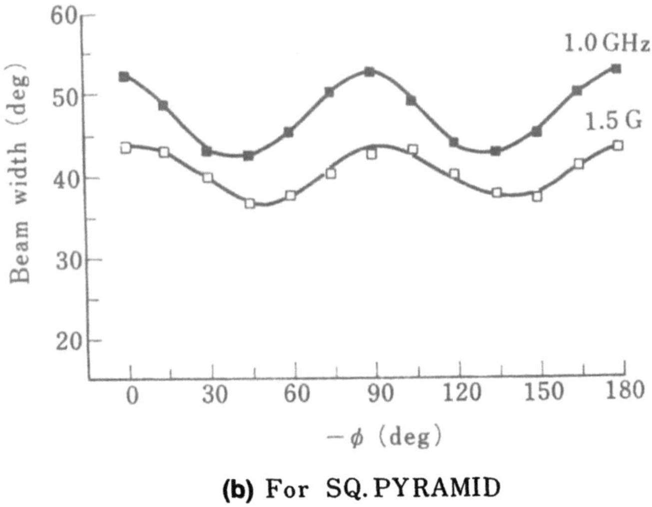

Fig. 9.12 Measured beam widths for varied ϕ: (a) Cone 1, (b) Cone 3.

9.4 Non-constant-impedance property of incorrectly arranged log-periodic structures

Often, there are serious misunderstandings that the log-periodic structure in itself will confer the constant-impedance property on antennas. It seems that such ignorance can be attributed to the term "log-periodic antenna" for the self-complementary log-periodic antenna, without reference to the most important fact that it is a derivative of the self-complementary structure [1.12, 1.13, 9.3].

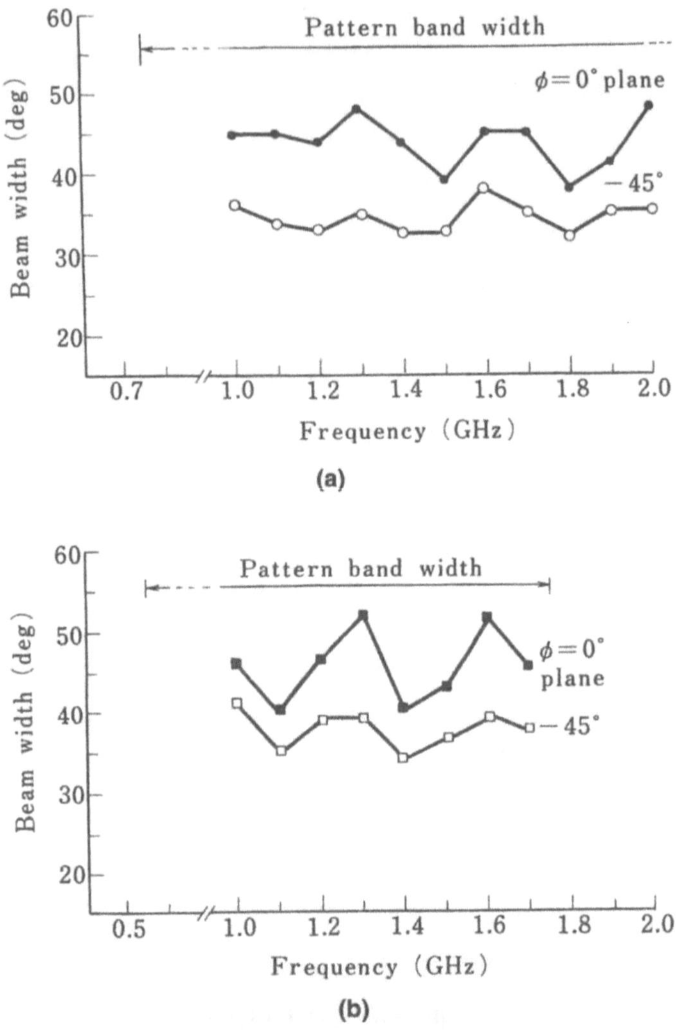

Fig. 9.13 Frequency characteristics of measured beam width: (a) Cone 1, (b) Cone 3.

In order to correct such misunderstandings, experimental tests have been done by taking a conically-bent modified SCA which is arranged in the log-periodic manner as shown in Fig. 9.15(a). As the most straightforward arrangement of non-self-complementary log-periodic structure, the antenna shown in Fig. 9.15(b) was constructed, where one wing of the two half-structures of the antenna is upside-down [1.6].

The measured values of input resistance for these two antennas are compared in Fig. 9.16, and a significant difference is apparent between them, in spite of the fact that the two wings of both antennas are identical. The input

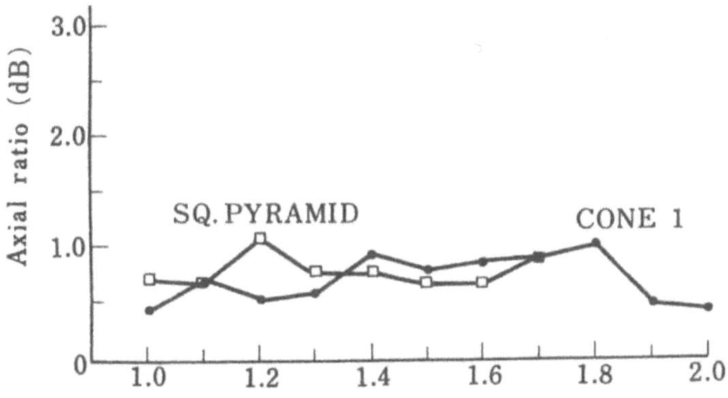

Fig. 9.14 Measured axial ratios for circularly polarized wave.

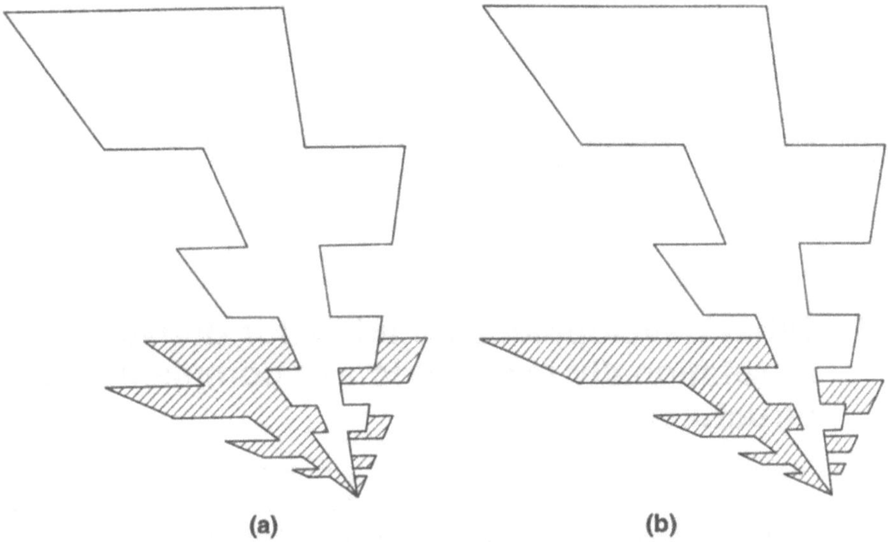

Fig. 9.15 Two kinds of log-periodic antenna: (a) log-periodic MSCA, (b) log-periodic antenna arranged in anti-complementary manner.

resistance for the incorrectly arranged log-periodic structure, as shown in Fig. 9.16(b), varies distinctly in a log-periodic manner for varying frequency, though the constant-resistance property is satisfactory for the modified SCA shown in Fig. 9.15(a).

From the results described above, it can be concluded that the origin of the broad-band property of the "log-periodic antenna" is not in its log-periodic shape, but rather in the aspect of the shape that is derived from the self-complementary structure.

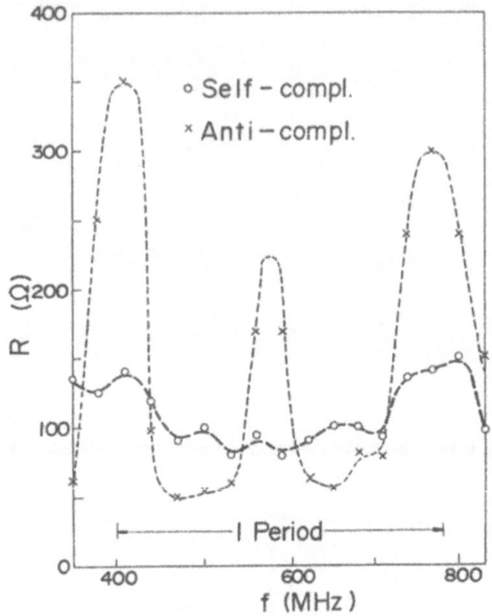

Fig. 9.16 Measured input resistances: ○ log-periodic MSCA; × log-periodic antenna, arranged in anti-complementary manner.

9.5 Other developmental studies for derivatives of rotationally symmetric self-complementary structures

Various developmental studies of rotationally symmetric self-complementary antennas, in addition to those discussed above, have been made incorporating approximations and modifications by means of deformations. Only those studies performed at Tohoku University and its related institutions are briefly described below.

The first is another alternate-leaves type SCA, where the shape of the leaves is in the form of quadratic curves [9.4] instead of linear strips or circular arcs.

The second is dipole arrays incorporating tapered distributions of the dipole lengths, where the directions of excitation are alternately interchanged for every adjacent dipole element. These are tapered dipole arrays with transposed excitation. Accordingly, if such a structure is arranged log-periodically, then it reduces to the log-periodic dipole array (LPDA) [1.13]. For this reason, this kind of dipole array can be considered as a modification of the LPDA.

Developmental studies in this category include the compound-spaced dipole array with distribution of dipole lengths tapered as a catenary curve [9.5], and the log-periodic dipole array with parasitic elements [9.6].

A completely different type of planar self-complementary antenna was proposed by N. Inagaki. This consists of a two-dimensional array of square conducting sheets arranged as a chequer-board [9.7]. When this structure is excited by periodic sources, it forms a rotationally symmetric self-complementary structure. It is called the "Ichimatsu moyou antenna".

10 • DEVELOPMENTAL STUDIES OF AXIALLY SYMMETRIC SELF-COMPLEMENTARY ANTENNAS

10.1 Experimental study of equally spaced unipole-notch array antenna

As an example of an axially symmetric self-complementary antenna for practical application, an equally spaced unipole (or monopole)-notch type array antenna with tapered distribution of the unipole-lengths as shown in Fig. 10.1 has been experimentally studied [8.1–8.3, 10.1–10.4]. According to the theory described in Chapter 4, this has a constant input impedance of $60\pi\ \Omega$ under the condition that the end terminals are terminated by a lumped resistance or a transmission line with $60\pi\ \Omega$. However, the end terminals of the tested structure are opened as shown in the same figure.

Measurements of the input impedance have been made in the frequency range of 1.0–2.0 GHz by varying the pitch of the teeth in the teeth-type array. Figures 10.2(a) and (b) show examples of the results for just two cases from the results obtained for S = 1, 2, 3 and 4 cm.

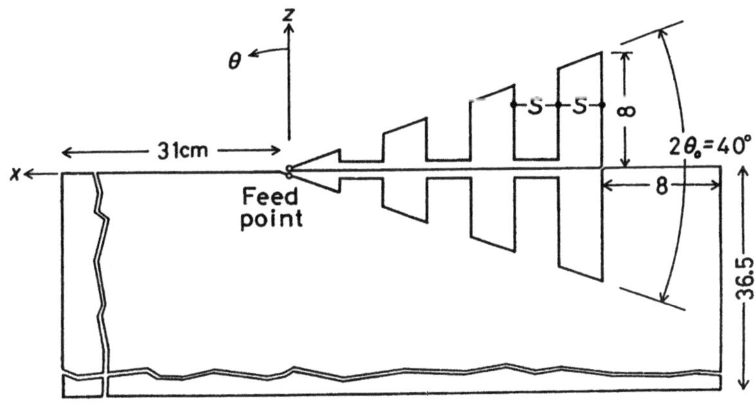

Fig. 10.1 Equally spaced axially symmetric SCA.

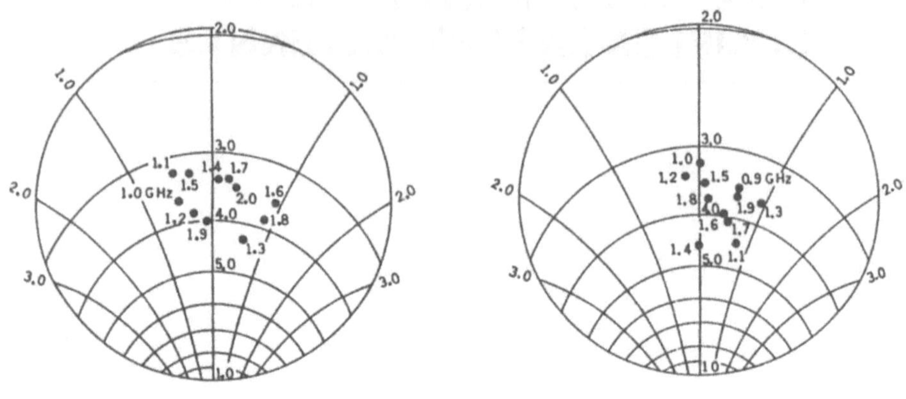

(a) (b)

Fig. 10.2 Measured input impedance of equally spaced SCA ($Z_0 = 50\ \Omega$): (a) S = 2 cm, (b) S = 4 cm.

The impedances are normalized to 50 Ω in these Smith charts. It is found from the charts that concentration of the values of impedance around the point $60\pi\ \Omega$ on the chart is satisfactory for all cases.

Measurements of the radiation characteristics of unipole-notch (U-N) type antennas have also been made by varying their various parameters. Examples of field patterns E_θ in the xz-plane are shown in Fig. 10.3. The cross-polarization

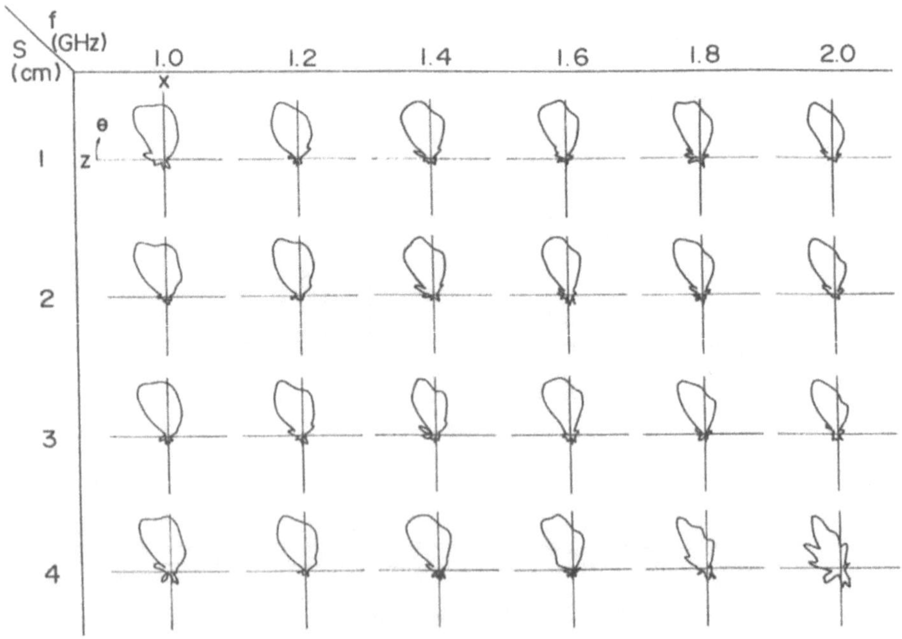

Fig. 10.3 Measured field patterns of equally spaced SCA ($2\theta = 40°$, $A = 31$, $B = 8$, $C = 36.5$ cm).

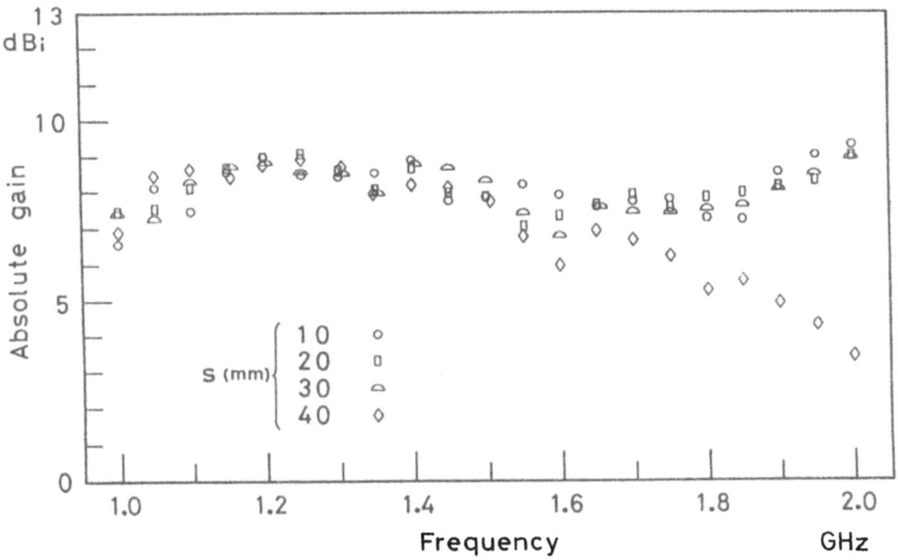

Fig. 10.4 Measured maximum gains of the equally spaced SCA for various frequencies.

component of the radiation field does not occur in the same plane as that which includes the antenna structure, but rather in the *xy*-plane as two side-lobes, one on each side of the *x*-axis.

According to the results of Fig. 10.3, uni-directional patterns remain almost unchanged for varying dimensions of *S* so long as it has small values. However, the uni-directional property deteriorates for antennas with increased dimensions of *S* near to a quarter-wavelength or larger. The direction of maximum radiation is tilted a little away from the *x*-axis, and slightly off the edge of the antenna sheet in the *xz*-plane.

Absolute gains in the direction of maximum radiation are plotted in Fig. 10.4. The results show that the curves of the gains are almost independent of the dimensions of *S* so long as it remains under a limit that is slightly smaller than a quarter-wavelength. The gradual slopes in the curves can be attributed to the effects of finiteness in the dimensions of the antenna sheet. The gain increases to some extent as the edge of the sheet in the *x*-direction becomes longer.

10.2 Equally spaced unipole-notch array antenna on a conical ground plane for the circularly polarized wave

As an application of the equally spaced U-N type array antenna for the circularly polarized wave, four structures of such an antenna are mounted on a conducting conical surface as illustrated in Fig. 10.5 [8.2, 8.3, 10.4].

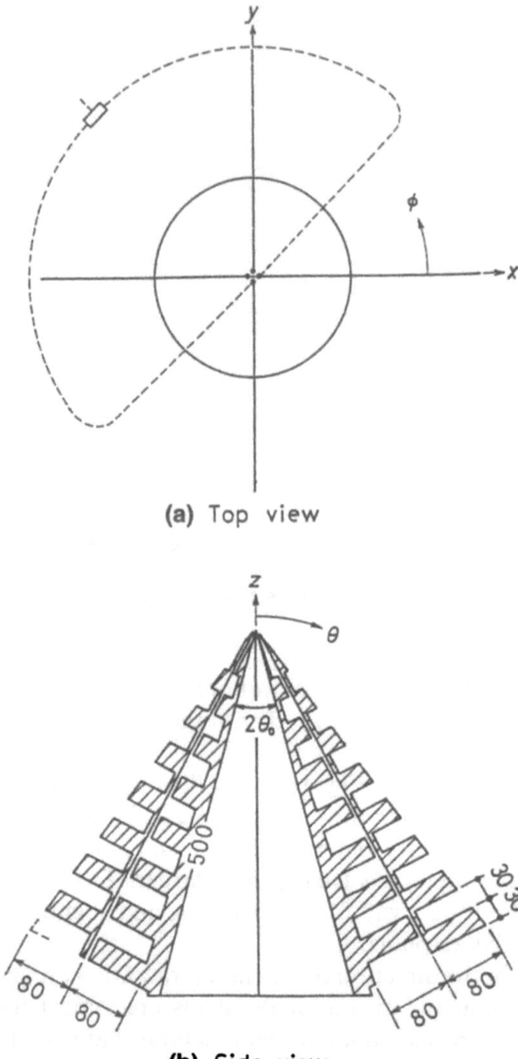

Fig. 10.5 Equally spaced axially symmetric SCA mounted on a conical surface: (a) top view, (b) side view.

In these structures, it should be noted that the actual shape of antennas mounted on opposite sides of the cone are different. Although they are almost identical, the phases of the periodic structures are shifted by a half-period so that the radiated fields are in opposite phase. For this reason, a pair of these antennas in alternate form can be excited in parallel connection, as shown in Fig. 10.5(a). This connection reduces the feeding impedance to half the original value, and the difference between the impedances of the antenna and that of the feeding cable can be reduced to a great extent.

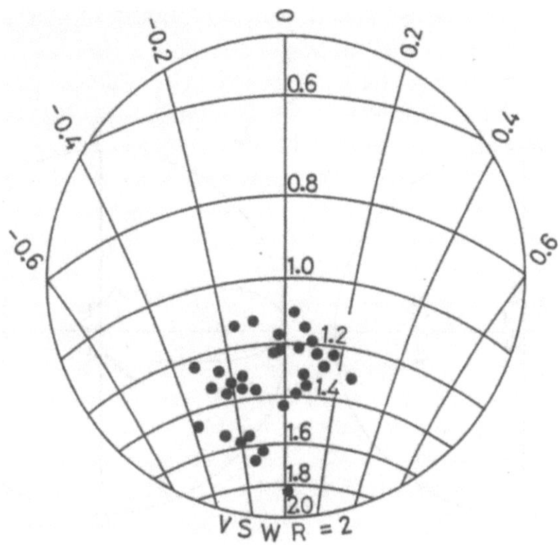

Fig. 10.6 Measured input impedances for one pair of SCA ($Z_0 = 50\ \Omega$, $2\theta = 30°$).

The measured input impedance of such a pair of antennas is plotted in Fig. 10.6. The concentration of the values of impedance around $30\pi\ \Omega$ is quite good for frequencies from 0.85 to 2.4 GHz. Figures 10.7(a) and (b) show the measured radiation patterns of E_θ in the xz-plane ($\phi = 0°$) and the yz-plane ($\phi = 90°$), while Fig. 10.8 shows the measured absolute gain for the same pair of antennas.

The cross-polarization components of the radiated fields have been found to be quite small.

Figure 10.9 shows the measured axial ratio for two pairs of array antennas which are excited with 90° phase difference as illustrated in Fig. 10.5(a).

The effects of the hybrid circuit are included in these results. Another feeding method which does not use the hybrid circuit has also been studied by applying the phase rotation principle [10.4].

10.3 Unipole-notch alternate array antennas stacked on both edges of an angular conducting sheet

As mentioned in section 10.1, the direction of maximum radiation of a U-N array antenna is slightly tilted from the direction of the symmetrical axis. By making use of this property, two U-N arrays stacked on both edges of an angular conducting sheet have been studied [10.3, 10.4]. This type of structure is conceivable also as a simplified deformation of a pair of arrays on a conical ground plane, as explained in section 10.2. Figure 10.10 shows two examples of

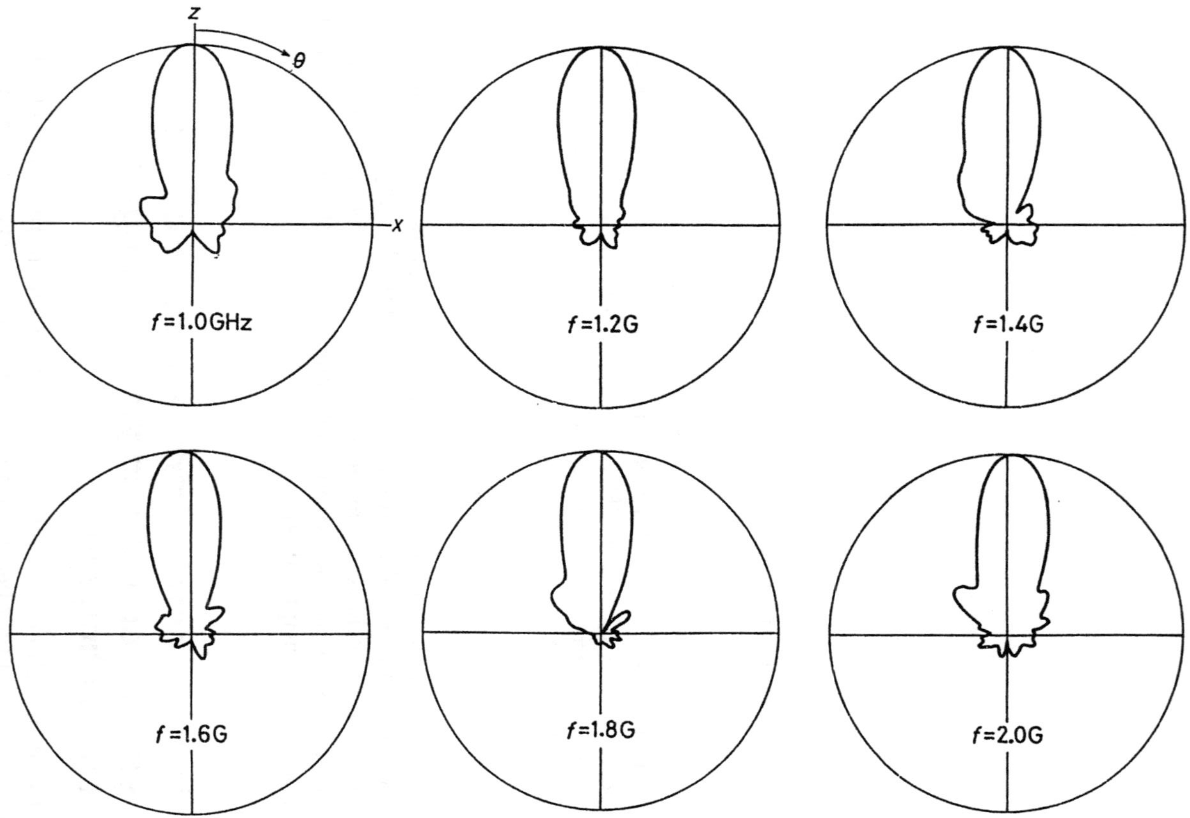

(a)

Fig. 10.7 Measured field patterns E_θ ($2\theta = 30°$): (a) *xz*-plane.

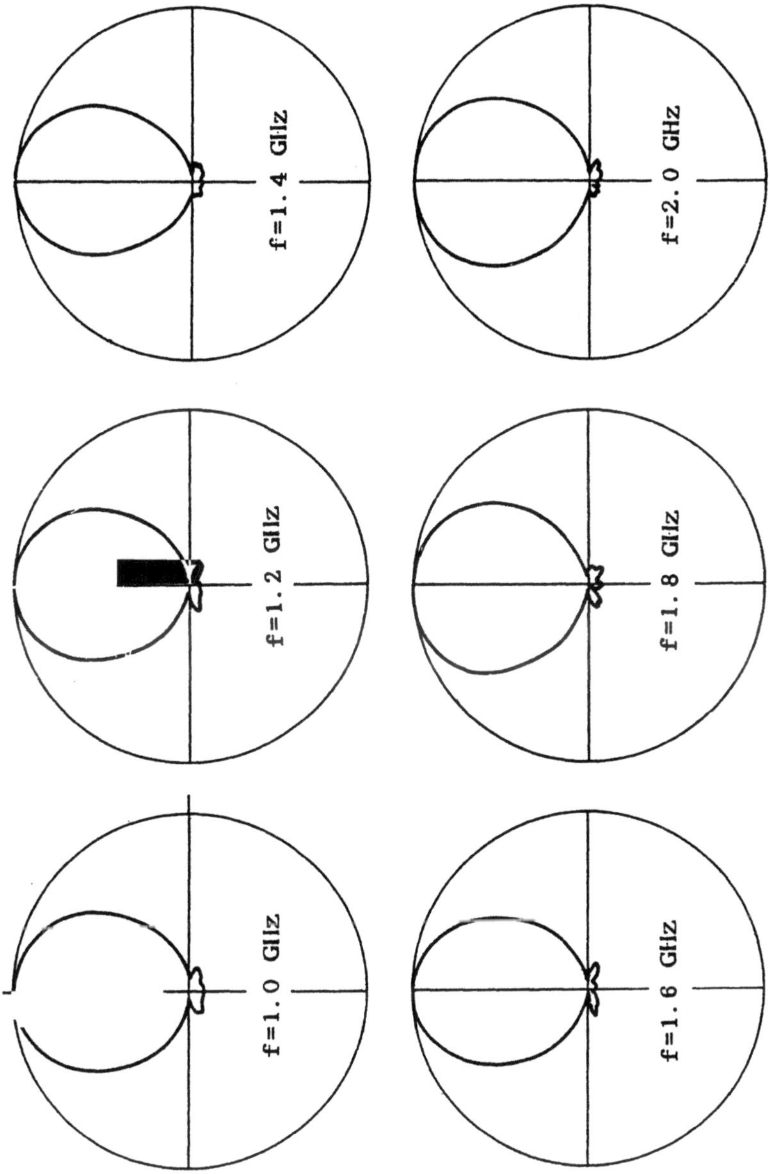

(b)

Fig. 10.7 (continued) (b) *yz*-plane.

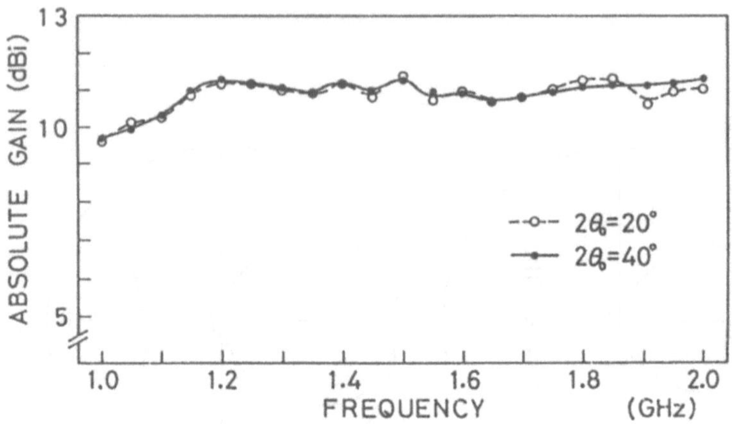

Fig. 10.8 Measured absolute gains for *xz*-polarization.

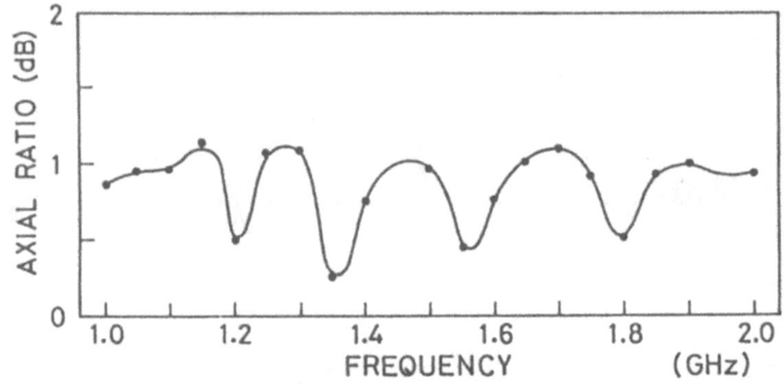

Fig. 10.9 Measured axial ratios for circularly polarized wave.

such a structure, where a pair of arrays are placed in alternate form, as shown in Fig. 10.5(b). Accordingly, these two array antennas can be excited in parallel connection as illustrated in Fig. 10.10(a). The structure in the same figure shows a proportionally spaced array, and Fig. 10.10(b) shows an equally spaced array.

The measured input impedances of these two antennas for frequencies from 0.8 to 2.0 GHz are normalized to 50 Ω and plotted in Figs 10.11(a) and (b). Concentrations of the values of impedance around 30π Ω on Smith charts are fairly good for both cases, in spite of the fact that the structures are drastically modified compared with the originals, which are placed on a semi-infinite planar conducting sheet.

The measured radiation patterns for these two antennas are fairly similar and remain unchanged for frequencies from 0.8 to 2.0 GHz. The beam widths in the *H*- or *xy*-plane are always broader than those in the *E*- or *xz*-plane. Examples of the field patterns E_θ for 1.0 GHz and 1.6 GHz are shown in Figs. 10.12(a) and (b).

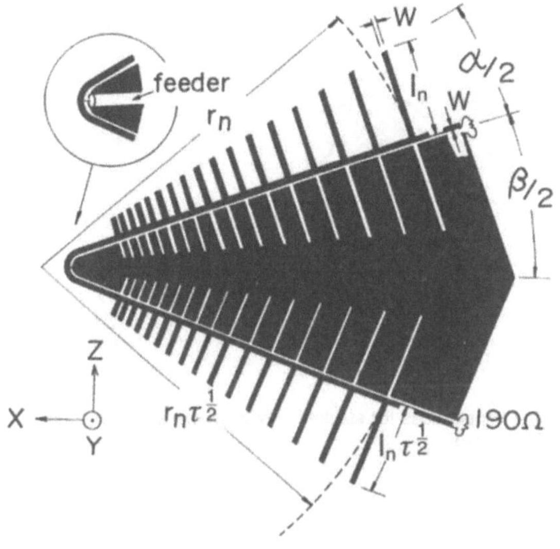

(a)

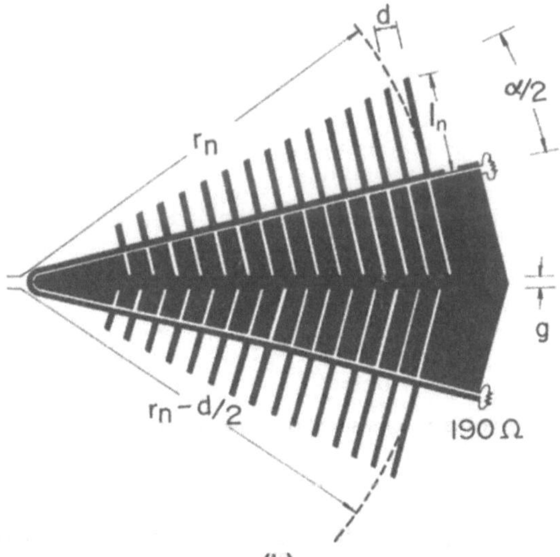

(b)

Fig. 10.10 Stacked U-N alternate array antennas: (a) proportionally spaced, $\alpha = 28°$, $\beta = 40°$, $\tau = r_i/r_{i-1} = l_i/l_{i-1} = 0.9$, $(r_i - r_{i-1})/4l_i = 0.1$, r_n = 36 cm, l_n = 9.2 cm, W = 4 mm; (b) equally spaced, $\alpha = 28°$, g = 1 cm, d = 2 cm, r_n = 36 cm, l_n = 9.2 cm, W = 4 mm.

The cross-polarization component, E_ϕ, of the radiation field does not exist in the xz-plane which includes the antenna structure, but a small amount of E_ϕ-component is observed in the xy-plane. Most of the cross-polarization component observed for the single array shown in Fig. 10.1 is suppressed in these stacked structures. The directive gains obtained for these antennas are more than 9 dBi.

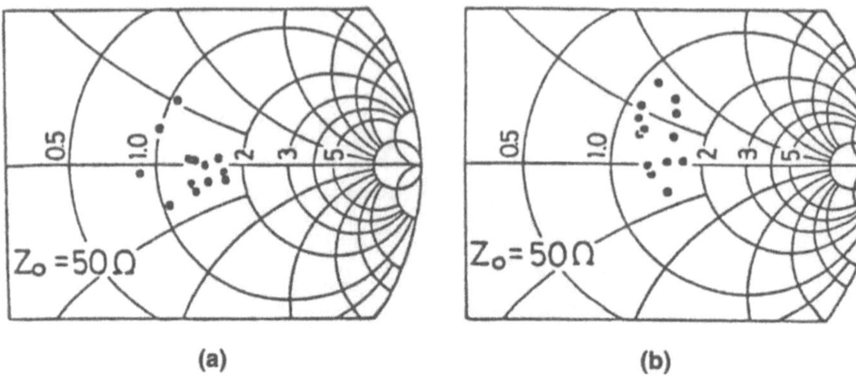

Fig. 10.11 Measured input impedances for stacked U-N alternate array antennas ($Z_0 = 50\ \Omega$): (a) proportionally spaced, (b) equally spaced.

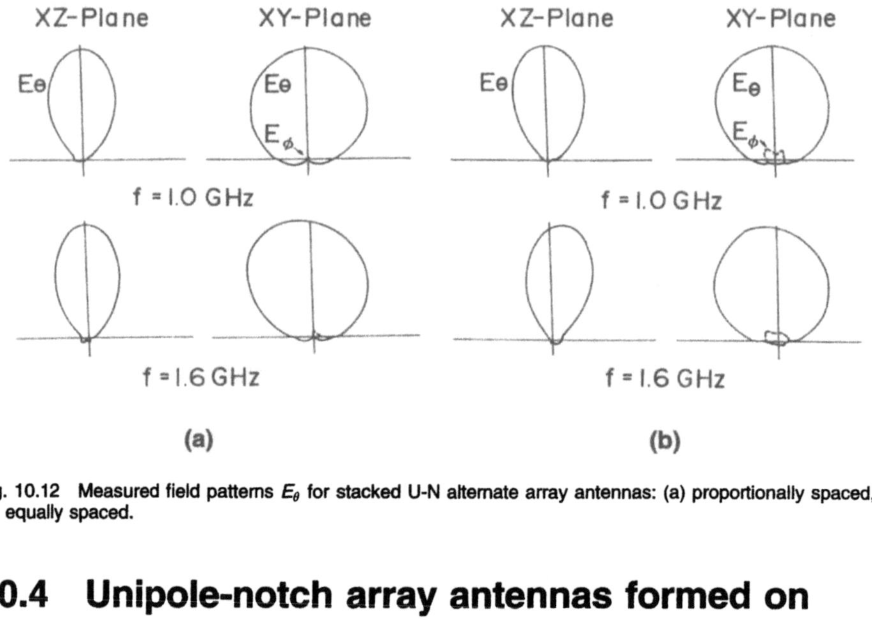

Fig. 10.12 Measured field patterns E_θ for stacked U-N alternate array antennas: (a) proportionally spaced, (b) equally spaced.

10.4 Unipole-notch array antennas formed on the substrate of a printed circuit

For the purpose of improving the mechanical strength of U-N array antennas, the substrate of a printed circuit has been introduced into these structures, and the effects of the dielectric plate have been studied from various points of view [8.3, 10.3, 10.5].

In these developmental studies, Teflon glass substrate of several thicknesses (for example, 0.9–2.0 mm) have been used in the experiments. The numerical results obtained from the effects of the dielectric plates depend on the thickness. For this reason, some of the important results or useful information are itemized below:

1 The wavelength and the characteristic impedance of the self-complementary transmission line both decrease on the substrate.
2 The resonant lengths of a slender unipole and a slender notch become shorter on the substrate.
3 The shortening percentage for the resonant length of the unipole on the substrate is a little smaller than that of the notch.
4 The input resistance at resonance decreases for the unipole, but increases for the notch.
5 U-N array antennas can be formed on the substrate of a printed circuit by reducing various dimensions compared with those in free space. The concentration of the input impedance on a Smith chart is satisfactory.
6 U-N alternate array antennas stacked on both edges of an angular conducting sheet can be formed on the substrate, without appreciable deterioration of the constant-impedance property and the radiation characteristics.

Some numerical values for an example of the substrate of a printed circuit are shown in Table 10.1, and numerical examples for U-N array antennas on that substrate are listed in Table 10.2.

These data have been obtained as the results of the developmental studies mentioned above.

Table 10.1 Example of numerical constants for a printed-circuit substrate

No.	Item	Numerical constant
1	Specific permittivity	$\epsilon_r = 2.6$
2	Loss tangent	$\tan\delta = 0.0022$
3	Intrinsic impedance	$Z_s = 233.6\,\Omega$
4	Wavelength reduction factor	62%
5	Thickness of the dielectric	1 mm

Table 10.2 Numerical examples obtained for U-N array antennas on a printed-circuit substrate* (0.8–7.0 GHz)

No.	Description	In free space (in dielectric)	On substrate
1	Wavelength along self-complementary transmission line	100% (62%)	77%
2	Resonant length of unipole (versus quarter-wavelength)	93–94%	80–81%
3	Resonant length of notch (versus quarter-wavelength)	93–94%	73–76%
4	Designed length of unipole (versus quarter-wavelength)	100%	88%
5	Designed ratio of notch length to unipole length	1	0.92–0.97
6	Input resistance of unipole at resonance	about 80 Ω	about 70 Ω
7	Input resistance of notch at resonance	about 440 Ω	450–500 Ω
8	Characteristic impedance of self-complementary transmission line	188 Ω (117 Ω)	about 145 Ω
9	Centre of concentration for input impedance	188 Ω (117 Ω)	about 145 Ω

*Widths of the self-complementary transmission line, unipole and notch: 1 mm.

11 • MONOPOLE-SLOT TYPE MODIFIED SELF-COMPLEMENTARY ANTENNAS

11.1 Monopole-slot type antennas derived from the three-dimensional self-complementary antenna

The radiation of the three-dimensional SCA shown in Fig. 6.7 is separated into two portions which are respectively in the upper and the lower half-spaces divided by the infinite horizontal plane of the conducting sheet. By taking this property into account, a practical application of this antenna is conceivable, where only the upper half of its radiation is utilized.

To realize such a practical antenna, a monopole-slot (M-S) type new antenna, as shown in Fig. 11.1 is proposed [8.8, 8.9, 11.1].

The top view of this antenna is identical to that of the three-dimensional SCA as shown in Fig. 6.7, but the lower half of the vertical structure is excised from the original structure. Accordingly, the radiation to the lower half-space from the new structure must be considerably reduced, though the radiation to the upper half-space is expected to be unchanged. However, it seems that a theoretical general treatment of such a structure is not straightforward. For this reason, the author and his colleagues have studied this type of modified SCA (MSCA) mainly by experiment.

Initially, the frequency characteristics of the input impedance of this antenna over the frequency range 0.95–2.2 GHz were studied. The measured results are plotted in Fig. 11.2, where the values of impedance are normalized at 50 Ω. The chart shows that the constant-impedance property is preserved quite well, in spite of a great deal of deformation from the self-complementary structure. However, the centre of the concentration is evidently shifted towards a higher value compared with the original impedance of 30π Ω. Incidentally, further discussions about the value of the input impedance of this type of structure will be made in the next section by considering some limiting cases.

The radiation characteristics of this antenna were also studied. The measured results show that the radiation patterns are almost independent of the frequencies in the range 1.0–2.0 MHz. An example of the pattern E_ϕ in the xy-plane is shown in Fig. 11.3.

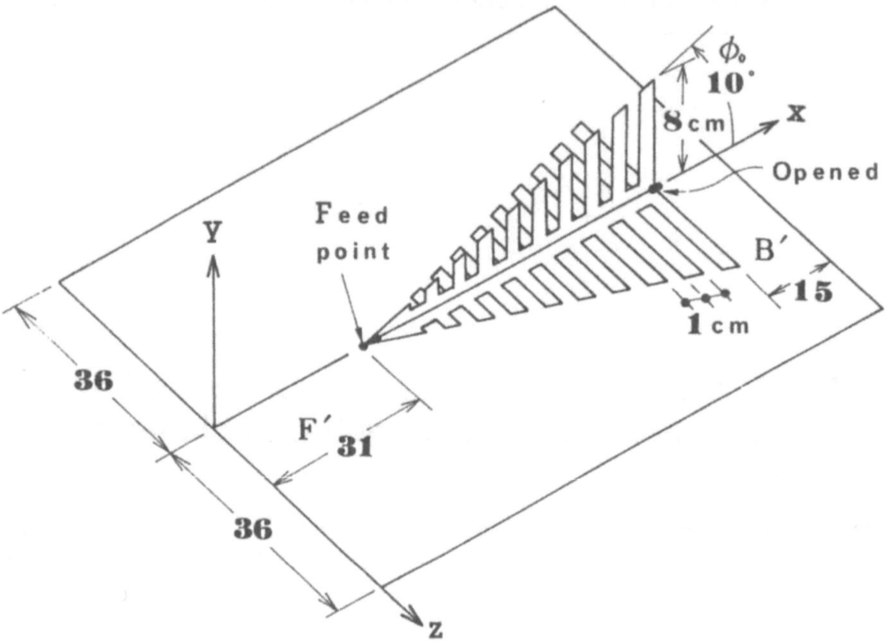

Fig. 11.1 M-S type MSCA.

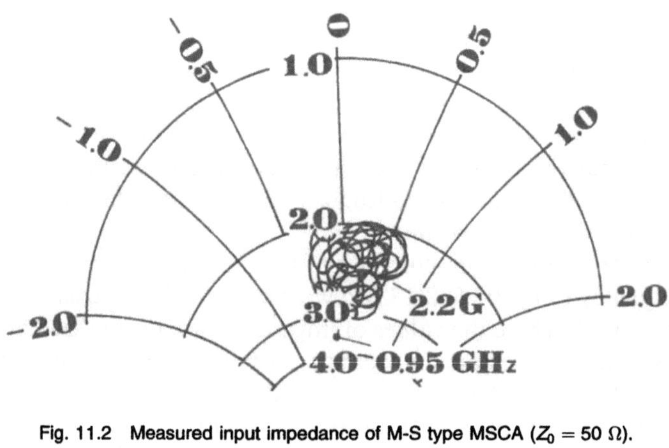

Fig. 11.2 Measured input impedance of M-S type MSCA ($Z_0 = 50\ \Omega$).

The elevation angles for the maximum radiation ϕ_{max} are around 25° or a little larger. The radiations to the lower half-space below the horizontal plane are at the level of about -5 dB or less, compared with the maximum radiation. Figure 11.4 shows an example of the measured field pattern in the plane which includes the direction of the maximum radiation and the z-axis.

The variations of ϕ_{max} and the maximum power gain for various values of parameters are shown in Figs 11.5 and 11.6.

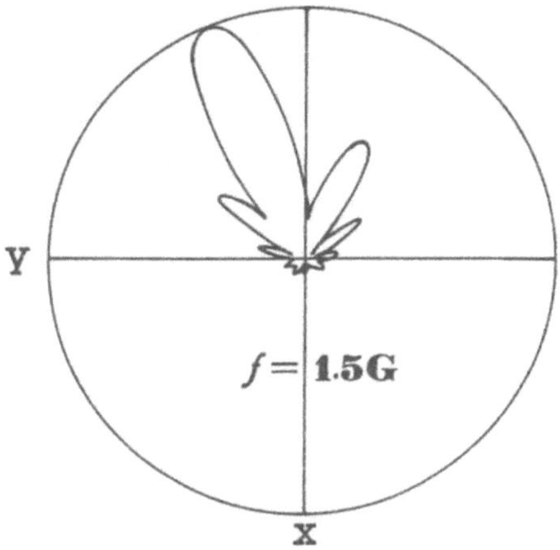

Fig. 11.3 Measured field pattern E_ϕ of M-S type MSCA in the *xy*-plane.

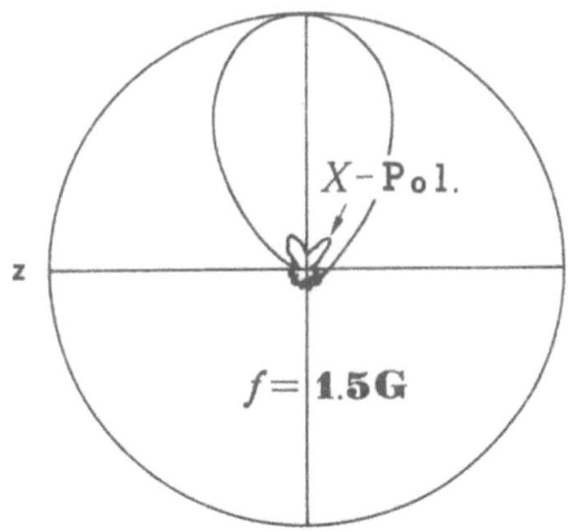

Fig. 11.4 Measured field pattern of M-S type MSCA in the plane that includes the direction of ϕ_{max} and the *z*-axis.

Further developmental studies on this type of MSCA have also been made from the view-point of practical applications [8.3, 8.7, 10.3, 10.4].

11.2 A monopole-slot antenna element as a limiting case

The most simplified structure of a monopole-slot type antenna is the slender monopole-slot antenna element illustrated in Fig. 11.7(a). This is equivalent to the three-dimensional SCA shown in Fig. 6.6. Such an antenna can be easily analysed by the conventional theories of the linear antenna and the slot antenna.

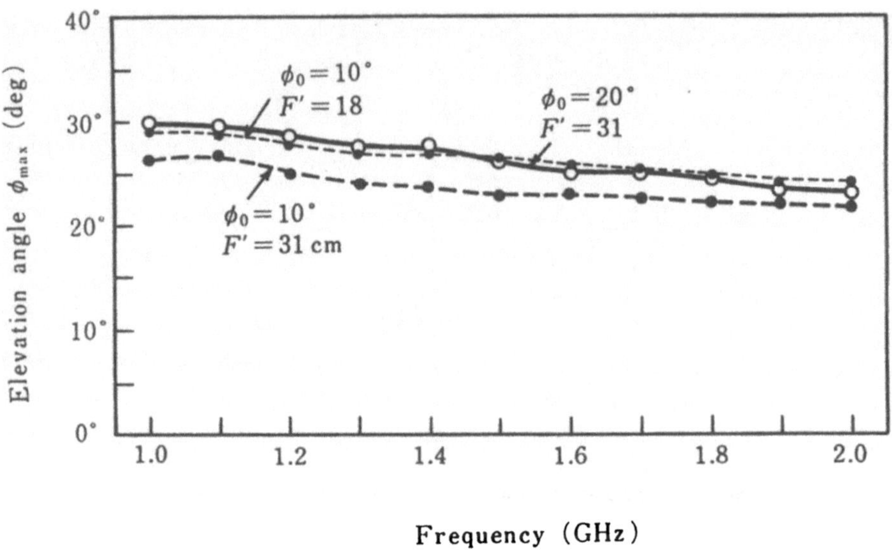

Fig. 11.5 Measured ϕ_{max} of M-S type MSCA.

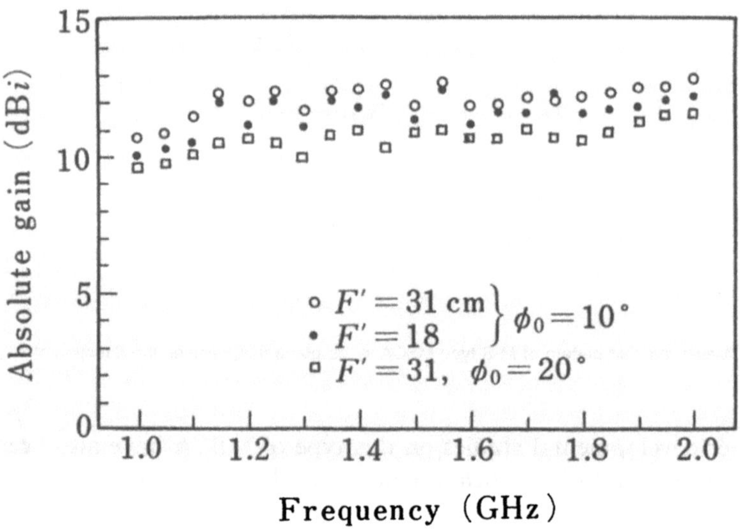

Fig. 11.6 Measured absolute gains of M-S type MSCA.

Let the voltages and currents at the two ports in Fig. 11.7(a) be V_1, I_1 and V_2, I_2, respectively, and those of the monopole antenna and the slot antenna be V_a, I_a and V_s, I_s, respectively.

Then, from the equivalent connection of the feeding circuits illustrated in Fig. 11.8, we obtain

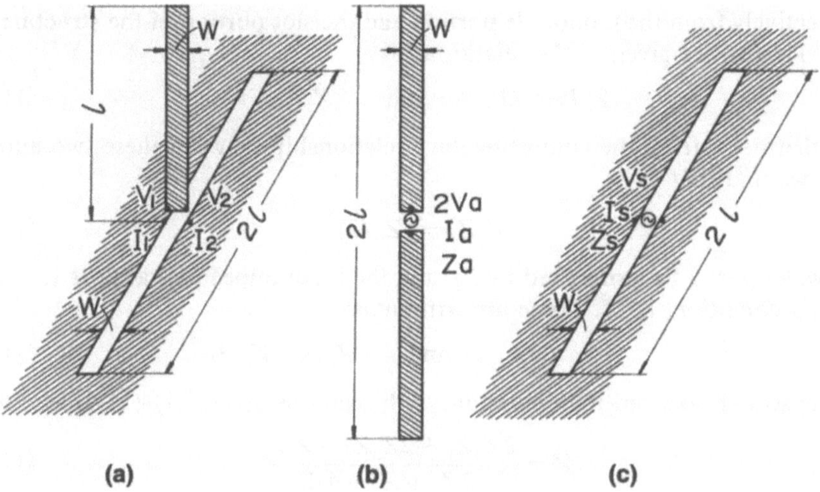

Fig. 11.7 Slender M-S antenna element and corresponding elementary antennas: (a) M-S element ($Z \simeq 30\sqrt{2}\pi\ \Omega$), (b) dipole antenna, (c) slot antenna.

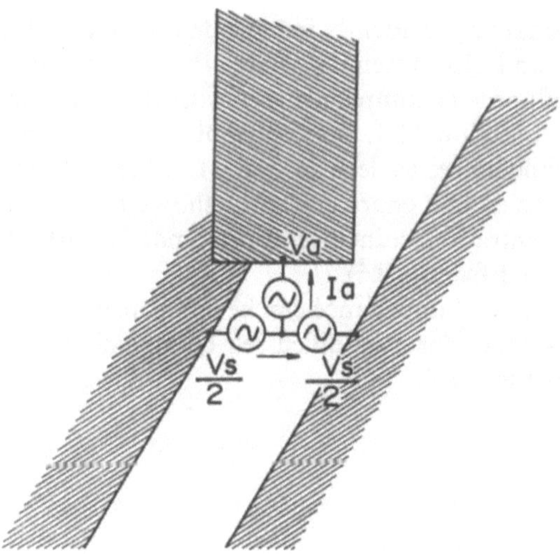

Fig. 11.8 Equivalent connection of feeding circuit.

$$\left.\begin{array}{ll} I_1 = (1/2)I_a + I_s & V_1 = V_a + (1/2)V_s \\ I_2 = (1/2)I_a - I_s & V_2 = V_a - (1/2)V_s \end{array}\right\} \qquad (11.1)$$

In addition, the input impedances Z_a and Z_s for the dipole antenna shown in Fig. 11.7(b) and the slot antenna shown in Fig. 11.7(c), which are derived respectively from the monopole portion and the slot portion of the structure in Fig. 11.7(a), are given in the relationships

$$Z_a I_a = 2V_a \qquad \text{and} \qquad Z_s I_s = V_s \tag{11.2}$$

Furthermore, from the complementary relationship between these two antennas, we find that

$$Z_a Z_s = (Z_0/2)^2 \tag{11.3}$$

Now, let port 2 be terminated by Z_l, and the input impedance at port 1, under such a condition, be Z. These are written as

$$Z = V_1/I_1 \qquad \text{and} \qquad Z_l = -V_2/I_2 \tag{11.4}$$

Some straightforward manipulations with relationships (11.1)–(11.4) lead to

$$Z = \frac{2Z_a Z_s + 2Z_a Z_l + Z_s Z_l}{2Z_a + Z_s + 4Z_l} \tag{11.5}$$

In this equation, if we assume that the input impedance is equal to the loaded impedance, then we obtain

$$Z = Z_l = Z_0/2\sqrt{2} \simeq 30\sqrt{2}\pi \qquad [\Omega] \tag{11.6}$$

This value of impedance is evidently independent of source frequency and sizes of the monopole and slot antennas. Here, the dipole antenna which corresponds to the monopole is complementary with the slot antenna.

As mentioned in section 11.1, the centre of the concentration for the measured values of impedance, as seen in Fig. 11.2, has not yet been determined theoretically. However, the chart evidently shows that the impedance at the centre of the concentration coincides approximately with $30\sqrt{2}\pi$ Ω which is theoretically obtained from (11.6).

This particular value of impedance can be confirmed by a limiting case of the characteristic impedance for a modified self-complementary transmission line, which is shown in Fig. 11.9.

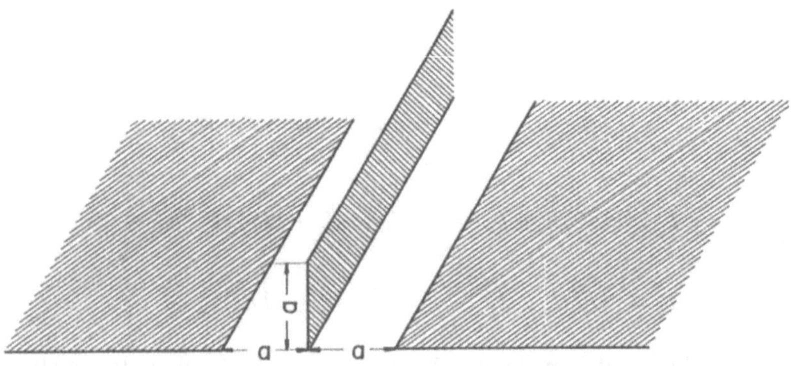

Fig. 11.9 Geometry of modified self-complementary transmission line ($Z \simeq 30\sqrt{2}\pi$ Ω).

This transmission line can be derived from the self-complementary transmission line in Fig. 6.5(a), by excising the lower half of the vertical structure. If we assume that the transverse cross-section of the new transmission line is sufficiently small compared with the wavelength, then the characteristic impedance Z_c can be calculated by the method of conformal mapping, as a limiting case where the electromagnetic fields are reduced to the static fields [10.3, 10.4]. Thus, the impedance was obtained by T. Ishizone as

$$Z_c \simeq 30\sqrt{2}\pi \ [\Omega] \tag{11.7}$$

This impedance is identical to the value obtained from (11.6).

These interesting theoretical results suggest that the experimentally obtained broad-band characteristics and the value of the input impedance for monopole-slot type antennas must have some theoretical foundation. However, such a new theory is still to be investigated, as well as other problems related to modified self-complementary structures.

12 • CONCLUSION

Self-complementary antennas have constant input impedance, independently of the source frequency and of the shapes of their structures. This nature has its origin in the "principle of self-complementarity", and various types of extremely broad-band antennas have been developed from this basic principle as its practical applications. In this book, such types of self-complementary antenna and their origin, the "principle of self-complementarity", are treated in detail.

Initially, the background to the emergence of these new types of antenna structure was reviewed, and a brief history of self-complementary antennas was chronologically traced, placing particular emphasis on the research performed by the author.

Next, the rigorous treatments of the fundamental theories that lead to the discovery of the "principle of self-complementarity" were made in detail, and the derivation process for self-complementary planar structures in basic form was clearly explained.

After this, extensions of the theory to various other cases were discussed, including multi-terminal planar structures, three-dimensional structures and periodically stacked structures. All the self-complementary structures derived from theory have the constant-impedance property, but unfortunately they contain infinitely extended conducting sheets.

To overcome the inconveniences that arise from the theory, various approximations and/or modifications of self-complementary structures have been made for practical applications. Some of the results obtained from the mainly experimental extensive studies performed at Tohoku University, and its related institutions, are explained in the latter part of the book. However, the descriptions are limited to only important results that are essential to the process of developmental investigations for extremely broad-band practical antennas.

It is made plain in the text that there are still certain problems to be investigated in connection with various shortcomings. But the theory nonetheless provides a firm foundation and effective technological guidance principle for investigations into extremely broad-band practical antennas. Furthermore, extensive applications of the "principle of self-complementarity", a significant and effective basic principle, may be expected not only for practical antennas, but also for various problems in other fields.

The author earnestly hopes for further developments and more extensive applications of this theory, both in pure science and in technological applications.

References

Chapter 1

1.1. Y. Mushiake, The input impedance of a slit antenna, *Joint Convention Record of Tohoku Sections of IEE and IECE of Japan*, pp. 25–26, June, 1948.
1.2. Y. Mushiake, The input impedances of slit antennas, *J. IEE Japan*, **69**, No. 3, pp. 87–88, March, 1949.
1.3. S. Uda and Y. Mushiake, The input impedances of slit antennas, *Tech. Rep. of Tohoku Univ.*, **14**, No. 1, pp. 46–59, September, 1949.
1.4. Y. Mushiake, Multiterminal constant impedance antenna, *1959 National Convention Record of IECE of Japan*, p. 89, October, 1959.
1.5. Y. Mushiake and H. Saito, Three-dimensional self-complementary antenna, *Joint Convention Record of Four Japanese Institutes Related to Electrical Engineering*, pt. 15, No. 1212, April, 1963.
1.6. Y. Mushiake, Constant impedance antennas, *J. IECE Japan*, **48**, No. 4, pp. 580–584, April, 1965.
1.7. Y. Mushiake, Self-complementary stacked antennas, *1982 Optical and Radio Division National Convention Record of IECE of Japan*, p. 90, August, 1982.
1.8. Y. Mushiake, Self-complementary antennas, *Researches on Elect. Comm., Record of Elect. Comm. Eng. Conversazione, Commemorative Issue, RIEC, Tohoku Univ.*, pp. 109–116, September, 1985.
1.9. Y. Mushiake, Compoundly stacked self-complementary antennas, *Memoirs of Tohoku Institute of Technology*, **10**, pp. 73–76, March, 1990.
1.10. V.H. Rumsey, Frequency independent antennas, *1957 IRE National Convention Record*, pt. 1, pp. 114–118, March, 1957.
1.11. V.H. Rumsey, *Frequency Independent Antennas*, Academic Press, New York, 1966.
1.12. R.H. Duhamel and D.E. Isbell, Broadband logarithmically periodic antenna structure, *1957 IRE National Convention Record*, pt. 1, pp. 119–128, March, 1957.
1.13. D.E. Isbell, Log periodic dipole arrays, *IRE Trans. Ant. Prop.*, **AP-8**, No. 3, pp. 260–267, May, 1960.
1.14. H. Yagi and S. Uda, Projector of the sharpest beam of electric waves, *Proc. of the Imperial Academy (of Japan)*, **2**, No. 2, pp. 49–52, February, 1926.
1.15. E. Hallén, Theoretical investigations into the transmitting and receiving qualities of antennae, *Nova Acta, Uppsala*, Ser. IV, **11**, No. 4, pp. 3–44, 1938.
1.16. J.A. Stratton, *Electromagnetic Theory*, McGraw-Hill, New York, 1941.
1.17. Y. Mushiake, An exact step-up impedance-ratio chart of a folded antenna, *IRE Trans. Ant. Prop.*, **AP-2**, No. 4, p. 163, October, 1954.
1.18. S. Uda and Y. Mushiake, *Yagi–Uda Antenna*, Maruzen (sole agent), Tokyo, 1954.
1.19. S. Adachi, J.R. McDaugal and Y. Mushiake, Super loop antenna, *Rep. RIEC, Tohoku Univ.*, **9**, No. 1, pp. 1–8, December, 1957.
1.20. S. Adachi and Y. Mushiake, Directive loop antennas, *Rep. RIEC, Tohoku Univ.*, **9**, No. 2, pp. 105–112, February, 1958.
1.21. G.A. Deschamps, Impedance properties of complementary multiterminal planar structures, *IRE Trans. Ant. Prop.*, **AP-7** (Special Supplement), pp. S371–S378, December, 1959.
1.22. Y. Mushiake, Self-complementary antennas, *IEEE Ant. Prop. Magazine*, **34**, No. 6, pp. 23–29, December, 1992.

Chapter 2

2.1. M. Ito, Properties of complex oscillating electromagnetic field and unified equation for electromagnetic oscillation. (Fundamental theory of oscillating electromagnetic field. I), *Collected Papers of IEE Japan*, **3**, pp. 135–144, April, 1942.
2.2. M. Kotani, Unpublished material, 1945.

Chapter 3

3.1. M. Ito, Unpublished material, 1945.
3.2. T. Matsumoto, Unpublished material, 1945.
3.3. Y. Asami, T. Matsumoto and S. Matsuura, Study on the slit aerials, *J. IEE Japan*, **67**, No. 9, pp. 150–153, September, 1947.
3.4. H. G. Booker, Slot aerials and their relation to complementary wire aerials, *Proc. IEE*, pt. IIIA, **90**, No. 4, pp. 620–629, April, 1946.
3.5. T. Yoneyama and Y. Mushiake, Input impedance and gain function of notch antenna, *Rep. RIEC, Tohoku Univ.*, **13**, No. 1, pp. 55–63, October, 1961.

Chapter 5

5.1. T. Ishizone and Y. Mushiake, A self-complementary antenna composed of unipole and notch antennas, *1977 International IEEE AP-S Symposium Digest*, 8–1, June, 1977.

Chapter 8

8.1. T. Kasahara and Y. Mushiake, Axially symmetric and equally spaced self-complementary array antenna, *Trans. IECE Japan*, **J65-B**, No. 3, pp. 338–339, March, 1982.
8.2. T. Kasahara and Y. Mushiake, Equally spaced monopole-notch array antenna for circularly polarized wave, *IEEE Trans. Ant. Prop.*, **AP-31**, No. 5, pp. 812–814, September, 1983.
8.3. Y. Mushiake, *Report of the studies on self-complementary antennas with modifications and approximations*, Grant-in-Aid for Scientific Research, Ministry of Education, Japan, Project No. 56890010 (1981–1983), 127 pages, March, 1983.
8.4. R.H. Duhamel and F.R. Ore, Logarithmically periodic antenna designs, *1958 IRE International Convention Record*, pt. I, pp. 139–151, March, 1958.
8.5. S.A. Schelkunoff, *Advanced Antenna Theory*, John Wiley & Sons, New York, 1952.
8.6. T. Kasahara and Y. Mushiake, Self-complementary ground-plane antennas in three dimensions, *Trans. IECE Japan*, **J65-B**, No. 9, pp. 1109–1116, September, 1982.
8.7. T. Kasahara, K. Sawaya and Y. Mushiake, Modified three-dimensional self-complementary array antenna over a finite ground plane, *Trans. IECE Japan*, **J66-B**, No. 1, pp. 40–47, January, 1983.
8.8. T. Ishizone, T. Kasahara and Y. Mushiake, Modified two planes self-complementary antenna, *1978 International Symposium on Antennas and Propagation, Japan, Summaries of Papers*, paper A-93, pp. 145–148, August, 1978.
8.9. T. Ishizone, Y. Yokoyama, S. Nishimura and Y. Mushiake, Unipole-slot array antenna, *Trans. IECE Japan*, **J66-B**, No. 3, pp. 281–288, March, 1983.
8.10. Y. Mushiake, Principle of log-periodic antenna (comments), *Broadcast Engineering*, **13**, No. 8, pp. 441–444, August, 1960.

Chapter 9

9.1. T. Furuya, T. Ishizone and Y. Mushiake, Alternate-leaves type self-complementary antenna and its application to high gain broad-band antennas, *IECE Technical Report*, **AP-77**, No. 43, pp. 35–40, July, 1977.
9.2. T. Kasahara and Y. Mushiake, Four-terminal circularly polarized self-complementary antenna, *Trans. IECE Japan*, **J65-B**, No. 8, pp. 981–988, August, 1982.
9.3. P.E. Mayes, Frequency-independent antennas and broad-band derivatives thereof, *Proc. IEEE*, **80**, pp. 103–112, January, 1992.
9.4. K. Matsumura and N. Nasu, Alternate leaves type self-complementary plane antenna, *Proc. International U.R.S.I.-Symposium 1980 on Electromagnetic Waves*, Munich, pp. 324A/1–4, August, 1980.

9.5. Mushiake Lab., Tohoku Univ. and Radio Res. Lab., Ministry of Post & Telecomm., Compound-spaced dipole array with element-lengths tapered in catenary curve, Unpublished material, 1984.

9.6. T. Ishizone, H. Ishikawa, S. Horiguchi and Y. Mushiake, Dipole array with log periodic parasitic elements, *1983 National Convention Record of IECE of Japan*, p. 716, April, 1983.

9.7. N. Inagaki, Y. Isogai and Y. Mushiake, Ichimatsu moyou antenna – self-complementary antenna with periodic feeding points, *Trans. IECE Japan*, **62-B**, No. 4, pp. 388–395, April, 1979.

Chapter 10

10.1. T. Ishizone and Y. Mushiake, Unipole-notch array antennas, *1981 International IEEE AP-S Symposium Digest*, S23-6, June, 1981.

10.2. K. Yamamoto, K. Sawaya, T. Ishizone and Y. Mushiake, Self-complementary monopole-notch array antennas, *Trans. IECE Japan*, **J65-B**, No. 1, pp. 70–77, January, 1982.

10.3. T. Ishizone, *Report of the studies on high gain and broad-band antennas derived from modified self-complementary antennas*, Grant-in-Aid for Scientic Research, Ministry of Education, Japan, Project No. 00555131 (1980–1982), 62 pages, March, 1982.

10.4. T. Kasahara, Studies on self-complementary broad-band antennas, Doctorate thesis submitted to Tohoku University, 172 pages, October, 1984.

10.5. S. Nishimura, T. Ishizone and Y. Mushiake, Input impedance of a monopole-notch antenna formed on a printed circuit substrate, *Joint Convention Record of Tohoku Sections of Institutes Related to Electrical Engineering*, 2B14, August, 1982.

Chapter 11

11.1. T. Kasahara and Y. Mushiake, Monopole-slot array antennas arranged on a conical surface, *1982 National Convention Record of IECE of Japan*, p. 620, March, 1982.

Subject Index

A

Absolute gain 101, 103, 106, 114
Alternate form 102, 106
Alternate-leaves type SCA 81–88
  log-periodic 86
  modified 88
Axial ratio 89, 95, 103, 106

B

Babinet's principle 3
Balanced-type 27, 28
Beam width 93, 94, 106

C

Catenary curve 96
Characteristic impedance 22, 46, 75, 109, 117
Compound-spaced 29, 96
Conformal mapping 117
Constant-impedance antenna 4, 27, 29
Constant-impedance property 27, 39, 40, 54, 57, 60, 72, 75, 81, 85, 87, 93, 109, 111, 119
Constant-resistance property 95
Coupling-less property 45–47
Cross-polarization component 85, 100, 107

D

Denki Kogyo Co. Ltd. 79
Directive gain 107
Duality relations 9

E

Equally spaced 29, 76, 81, 82, 99, 101, 102, 106, 107
Equivalent radius 21, 72, 75

F

Frequency-independent 1, 4, 79
Front-to-back ratio 87

G

Grant-in-Aid for Sci. Res. 4

H

Hallén's theory 2
Hertzean vector 10
Hole antenna 15, 16, 17
Hoso Bunka Foundation 4

I

Ichimatsu moyou antenna 97
Intrinsic impedance 16, 109

L

Log-periodic,
  antenna 1, 4, 75, 77, 79, 93, 95, 96
  dipole array (LPDA) 1, 4, 75, 79, 96
  incorrectly arranged 93–96
  non-self-complementary 94
  properly arranged 2
  self-complementary 77, 79, 93

M

Modified SCA (MSCA) 75, 88, 94, 111
Monopole antenna 22, 114
Monopole-slot (M-S) antenna 54, 113, 115
M-S type antenna 111, 112, 117
Mushiake Laboratory 4
Mushiake's relationship 27
Mutual admittance 18
Mutual impedance 18
Mutually complementary 13, 15
Mutually dual structure 9

N

Non-constant-impedance property 93–96
Notch antenna 22, 23

P

Parasitic element 96
Phase rotation principle 103
Plate antenna 15, 16
Power gain 112
Principle of self-complementary structure (Principle of self-complement) 3, 27, 119
Printed-circuit substrate (substrate of printed circuit) 108, 109
Proportionally spaced 29, 88, 89, 106, 107
Pseudo-self-complementary 62, 63

Q

Quarter-wave frequency 82, 85

R

Ring connection 31, 32, 33, 34

S

Self-admittance 18
Self-complementary antenna (SCA) 1, 5, 27, 28, 29
Self-complementary (planar) structure 1, 25, 26, 27
Self-impedance 18
Single-phase excitation 35, 37
Slot (slit) antenna 2, 3, 21, 22, 113, 114, 115, 116
Star connection 31, 32, 33, 34, 35, 37
Strip antenna 21, 22
Substrate of printed circuit (printed-circuit substrate) 108, 109

T

Tandem connection 47
Teeth-type 2, 72, 82, 87, 88, 99
Teflon glass substrate 108
Transmission line 22, 29, 46, 99, 117
  self-complementary 29, 30, 54, 109, 117
  modified self-complementary 116
Transposed excitation 79, 96
Transposition 1, 79
Truncation effects (effects of truncation) 2, 72, 87
Turnstile antenna 31

U

Unbalanced-type 29
Unipole antenna 22, 23
Unipole-notch (U-N) antenna 58, 63
U-N alternate array antenna 103–109
U-N array antenna 103, 108, 109
U-N type antenna 42, 47, 100
U-N type array antenna 99, 101

Y

Yagi-Uda antenna 2, 3

Author Index

Reference numbers are also listed.

A

Adachi S. 1.19, 1.20
Asami Y. 3.3

B

Booker G. 3.4

D

Deschamps G. A. 4, 37, 1.21
Duhamel R. H. 4, 1.12, 8.4

F

Furuya T. 9.1

H

Hallén E. 1.15
Horiguchi S. 9.6

I

Inagaki N. 4, 97, 9.7
Ishikawa H. 9.6
Ishizone T. 4, 117, 5.1, 8.8, 8.9, 9.1, 9.6, 10.1–3, 10.5
Isogai Y. 9.7
Isbell D. H. 4, 1.12, 1.13
Ito M. 2.1, 3.1

K

Kasahara T. 4, 47, 8.1, 8.2, 8.6–8, 9.2, 10.4, 11.1
Kotani M. 2.2

M

Matsumoto T. 3.2, 3.3
Matsumura K. 9.4
Matsuura S. 3.3
Mayes P. E. 9.3
McDaugal J. R. 1.19
Mushiake Y. 1, 5, 1.1–9, 1.17–20, 1.22, 3.5, 5.1, 8.1–3, 8.6–10, 9.1, 9.2, 9.6, 9.7, 10.1, 10.2, 10.5, 11.1
Mushiake Lab. 9.5

N

Nasu N. 4, 9.4
Nishimura S. 8.9, 10.5

O

Ore F. R. 8.4

R

Radio Res. Lab. 9.5
Rumsey V. H. 1, 4, 1.10, 1.11

S

Saito H. 1.5
Sawaya K. 4, 8.7, 10.2
Schelkunoff S. A. 8.5
Strattor J. A. 3, 1.16

U

Uda S. 2, 1.3, 1.14, 1.18

Y

Yagi H. 2, 1.14
Yamamoto K. 10.2
Yokoyama Y. 8.9
Yoneyama T. 3.5

MIX
Papier aus verantwortungsvollen Quellen
Paper from responsible sources
FSC® C105338

If you have any concerns about our products,
you can contact us on
ProductSafety@springernature.com

In case Publisher is established outside the EU,
the EU authorized representative is:
Springer Nature Customer Service Center GmbH
Europaplatz 3, 69115 Heidelberg, Germany

Printed by Libri Plureos GmbH
in Hamburg, Germany